How to Use this B

Use these math tests as a pre-assessment or post- asses
These math tests can be administered as a whole group, small group or when working individually with students.

Teacher Tips

- Review instructions with students to ensure understanding of the questions.
- Encourage students to complete the questions they know how to do first.
- Provide math manipulatives and other concrete materials for students to use during testing.

Math Rubric

Use the math rubric to encourage students to take responsibility for their learning and to assess their own math work.

Math Tests Record

This black line master will assist in keeping records.

Anecdotal Observations

Anecdotal observations provide good insight into a student's understanding and ability to apply the math concepts.

Table of Contents

Number Sense

And

Numeration Tests

Name ______________

Math Test: Addition and Subtraction

A. Add or subtract the following.

1. 909 - 176	2. 242 - 140	3. 329 + 583	4. 833 - 475	5. 315 + 338
6. 127 + 405	7. 435 - 127	8. 156 + 374	9. 184 + 275	10. 514 - 173

B. Solve the following problems.

	Show your work.
1. Sandy had 456 stamps. She gave 176 stamps to her little sister. How many stamps did she have left? There were ________ stamps left.	
2. There were 127 girls and 98 boys playing in the school yard. How many children were playing altogether? _________ children were playing.	

C. Circle the missing sign.

235 ◯ 169 = 404	+ — =

Name _______________

Math Test: Addition and Subtraction

1. 67 + 6

2. 49 + 18

3. 31 - 18

4. 63 - 28

5. 134 + 537

6. 566 + 174

7. 975 - 347

8. 409 - 158

9. 7986 + 1647

10. 45 967 + 12 563

11. 7542 - 2578

12. 7472 - 5175

13. 564 + 275 + 411

14. 3756 + 1036 + 5014

15. 89 407 - 69 758

16. 49 850 - 25 271

17. 65 331 + 33 986 + 13 745

18. 7356 + 1757 + 3611 + 7564

19. 66 108 - 11 977

20. 97 755 - 31 109

Number Correct

____
20

Name ________________

Math Test: Addition

4986	1785	7087	3082	7463	1337	1037
+ 1780	+ 4645	+ 1788	+ 5198	+ 6117	+ 7234	+ 1235
8523	2859	1763	3238	6379	8523	1523
+ 1370	+ 4752	+ 2097	+ 1634	+ 2717	+ 1470	+ 2692
4992	2611	1532	1754	8975	2034	
+ 3512	+ 1643	+ 8623	+ 2031	+ 1478	+ 3936	

Number Correct:

____
20

Math Test: Addition

Name ________________

5673	2073	1734	7081	1267	7745	2043
+ 5190	+ 2750	+ 3749	+ 1747	+ 7234	+ 6512	+ 1780
3092	8893	2875	1163	3892	6124	8869
+ 1906	+ 1310	+ 4832	+ 2097	+ 2450	+ 2613	+ 1312
1651	4992	4611	1402	1035	8705	
+ 5071	+ 34601	+ 1203	+ 8784	+ 3126	+ 1368	

Number Correct:

____
20

Name ________________

Math Test: Subtraction

7852 - 1378	7785 - 4049	7089 - 1788	8082 - 4136	7261 - 6017	6370 - 2234	5037 - 1563
8523 - 5630	2859 - 1787	5061 - 2413	8953 - 8624	3757 - 1718	8523 - 4988	2341 - 1558
4731 - 1312	4521 - 1784	9532 - 7431	2044 - 1754	8789 - 1404	5034 - 1936	Number Correct: ____ 20

Name ________________

Math Test: Subtraction

5980 - 2397	4723 - 2140	3734 - 1749	7634 - 1727	7288 - 1256	8740 - 5641	2093 - 1675
3092 - 1452	8893 - 1768	6785 - 4832	2007 - 1226	3892 - 1450	6865 - 2618	8162 - 1329
5651 - 1081	4902 - 1460	5871 - 1203	9578 - 1784	9035 - 3156	8705 - 3785	Number Correct: ____ 20

Name _______________

Math Test: Addition and Subtraction of Money

$134.50	$173.50	$45.80	$56.20	$94.87	$97.71	$56.75
- $ 70.45	- $ 90.72	- $21.06	- $10.09	- $32.38	- $53.33	- $13.37

$8.01	$29.08	$86.50	$84.53	$51.74	$4.70	$5.70
- $3.30	+ $65.75	- $43.03	- $33.06	+ $32.38	+ $3.37	+ $9.38

$51.70	$29.08	$24.83	$64.83	$54.72	$68.73
+ $15.73	+ $13.56	- $11.09	- $11.37	- $27.03	- $22.04

Number Correct: ____ / 20

Name _______________

Math Test: Addition and Subtraction of Money

$193.70	$50.82	$98.45	$154.76	$23.70	$91.75	$77.80
- $ 96.79	- $16.09	- $13.38	- $ 13.40	- $11.66	- $34.38	- $63.90

$89.00	$75.83	$5.70	$8.35	$81.23	$9.10	$7.12
+ $05.75	- $23.06	- $3.35	- $3.30	- $73.04	+ $9.38	+ $3.99

$69.08	$23.83	$38.73	$32.99	$44.22	$38.70
+ $10.53	- $11.37	- $27.03	+ $17.80	-$11.69	- $21.03

Number Correct: ____ / 20

Name _______________

Math Test: Multiplication Facts 1-10 #1

2	5	4	8	6	10	7
X 5	X 8	X 2	X 1	X 3	x 4	X 6
6	9	6	3	7	9	9
X 4	X 5	X10	X 2	X 4	X 1	X 6
1	10	8	7	10	10	Number Correct:
X 6	X 7	X 8	X 5	X 2	X 9	___ / 20

Name _______________

Math Test: Multiplication Facts 1-10 #2

3	6	5	4	7	8	4
X 5	X 1	X 10	X 3	X 7	X 2	X 4
8	9	7	1	10	2	3
X 4	X 7	X 3	X 4	X 5	X 8	X 9
4	6	5	8	3	3	Number Correct:
X 7	X 6	X 5	X 3	X 3	X 6	___ / 20

Name ______________

Math Test: Multiplication

69	73	92	89	84	13	41
x 83	x 25	x 86	x 25	x 63	x 45	x 25

39	35	78	17	22	40	97
x 56	x 64	x 22	x 64	x 27	x 48	x 64

33	54	42	88	18	60
x 14	x 17	x 56	x 17	x 35	x 32

Number Correct: ____ / 20

Name ______________

Math Test: Multiplication

74	49	37	60	28	49	19
x 93	x 35	x 56	x 73	x 15	x 25	x 35

75	45	68	96	23	80	45
x 36	x 74	x 42	x 61	x 48	x 84	x 94

63	82	38	55	80	82
x 17	x 42	x 28	x 39	x 45	x 76

Number Correct: ____ / 20

Name ______________

Math Test: Multiplication

34 x 13	85 x 35	72 x 56	63 x 45	89 x 66	45 x 75	77 x 25
31 x 46	38 x 54	76 X32	15 x 84	21 x 17	42 x 48	93 x 64
23 x 15	64 x 27	41 x 46	98 x 16	15 x 25	50 x 22	Number Correct: ___ / 20

Name ______________

Math Test: Multiplication

94 x 83	48 x 34	47 x 66	61 x 74	38 x 75	48 x 35	29 x 38
85 x 37	35 x 77	64 x 62	86 x 62	13 x 98	81 x 44	49 x 54
63 x 10	72 x 46	39 x 78	51 x 59	10 x 46	85 x 26	Number Correct: ___ / 20

Name ______________

Math Test: Multiplication

1. 20 X 50 =
2. 20 X 30 =
3. 50 X 50 =
4. 50 X 40 =
5. 40 X 10 =
6. 80 X 100 =
7. 70 X 20 =
8. 50 X 60 =
9. 40 X 70 =
10. 90 X 80 =
11. 70 X 50 =
12. 50 X 20 =
13. 60 X 60 =
14. 80 X 20 =
15. 20 X 60 =

Number Correct: /15

Math Test: Multiplication

Name ____________________

1. 10 X 10 =
2. 34 X 10 =
3. 6 X 10 =
4. 3 X 1000 =
5. 69 X 10 =
6. 70 X 100 =
7. 40 X 10 =
8. 22 X 10 =
9. 90 X 100 =
10. 34 X 10 =
11. 4 X 1000 =
12. 25 X 100 =
13. 22 X 100 =
14. 95 X 100 =
15. 100 X 10 =

Number Correct: /15

Name ______________

Math Test: Multiplication With Decimals

1.	2.	3.	4.	5.
2.3 x 5	3.8 x 8	6.9 x 3	1.2 x 4	7.3 x 5

6.	7.	8.	9.	10.
8.6 x 7	6.8 x 8	4.9 x 9	2.4 x 3	3.7 x 2

Number Correct: /10

Name ______________

Math Test: Multiplication With Decimals

1.	2.	3.	4..	5.
5.3 x 4.5	7.2 x 6.3	6.1 x 3.2	3.7 x 2.2	8.4 x 8.5

6.	7.	8.	9.	10.
9.3 x 2.1	1.8 x 4.9	2.5 x 8.7	7.4 x 1.6	8.2 x 3.3

Number Correct: /10

Name ________________

Division Facts 6-10 #1

36 ÷ 4 = 21 ÷ 3 = 36 ÷ 9 = 27 ÷ 3 = 10 ÷ 2 = 30 ÷ 5 = 24 ÷ 4 =

24 ÷ 6 = 18 ÷ 9 = 32 ÷ 8 = 35 ÷ 7 = 32 ÷ 4 = 48 ÷ 6 = 49 ÷ 7 =

64 ÷ 8 = 63 ÷ 9 = 30 ÷ 6 = 40 ÷10 = 14 ÷ 7 = 27 ÷ 9 =

Number Correct:

_____
20

Division Facts 6-10 #2

Name ________________

27 ÷ 9 = 6 ÷ 6 = 20 ÷ 10 = 90 ÷ 10 = 42 ÷ 7 = 40 ÷ 5 = 30 ÷ 10 =

45 ÷ 5 = 16 ÷ 2 = 5 ÷ 5 = 8 ÷ 2 = 12 ÷ 4 = 1 ÷ 1 = 30 ÷ 5 =

80 ÷ 8 = 54 ÷ 6 = 36 ÷ 9 = 72 ÷ 9 = 28 ÷ 7 = 40 ÷10 =

Number Correct:

_____
20

Math Test: Dividing Decimals

Name ______________

1) 2.6 ÷ 10 =	2) 5.4÷ 100 =
3) 716.9 ÷ 100 =	4) 62.9 ÷ 10 =
5) 426.9 ÷ 10 =	6) 985.2 ÷ 100 =
7) 0.3 ÷ 10 =	8) 62.4 ÷ 10 =
9) 593.1 ÷ 100 =	10) 204.6 ÷ 100 =

Number Correct: /10

Math Test: Dividing Decimals

Name ______________

1) 15.7÷ 100 =	2) 7.4 ÷ 100 =
3) 32.5 ÷ 10 =	4) 67.1 ÷ 10 =
5) 641.2 ÷ 100 =	6) 891.9 ÷ 100 =
7) 34.4 ÷ 10 =	8) 0.8 ÷ 10 =
9) 786.2 ÷ 10 =	10) 412.5 ÷ 100 =

Number Correct: /10

Name _______________

Math Test: Division- No Remainders

1)	2)	3)	4)	5)
5) 120	6) 534	5) 955	6) 420	3) 891

6)	7)	8)	9)	10)
8) 760	3) 798	9) 414	8) 720	5) 930

Math Test: Division- No Remainders

Name _______________

1)	2)	3)	4)	5)
7) 266	4) 900	2) 980	3) 342	7) 553

6)	7)	8)	9)	10)
8) 896	6) 468	9) 414	5) 830	3) 408

Name _______________

Math Test: Division- With Remainders

1) 5)621

2) 9)778

3) 4)352

4) 4)141

5) 6)609

6) 8)378

7) 7)925

8) 3)496

9) 8)483

10) 7)608

Name _______________

Math Test: Division- With Remainders

1) 5)531

2) 6)835

3) 5)577

4) 3)322

5) 9)250

6) 5)133

7) 6)182

8) 9)353

9) 9)784

10) 3)307

Math Test: Division- No Remainders

Name ______________

1)	2)	3)	4)	5)
90) 7470	50) 100	30) 750	20) 3400	80)400

6)	7)	8)	9)	10)
30) 150	90) 4950	20)4160	30) 7860	70)840

Math Test: Division- With Remainders

Name ______________

1)	2)	3)	4)	5)
15) 8956	96) 4492	18) 6865	11)8858	65) 6557

6)	7)	8)	9)	10)
46)1439	51) 5821	16)1270	81)820	13)3712

Name ________________

Math Test: Division with Decimals

1) $5\overline{)4.0}$

2) $6\overline{)2.1}$

3) $6\overline{)4.38}$

4) $9\overline{)6.3}$

5) $8\overline{)5.400}$

6) $6\overline{)2.358}$

7) $2\overline{)8.24}$

8) $2\overline{)4.58}$

9) $1.6\overline{)5.28}$

10) $7.8\overline{)7.02}$

11) $4.9\overline{)1.47}$

12) $2.6\overline{)8.32}$

13) $3.4\overline{)15.3}$

14) $1.4\overline{)8.82}$

15) $7.4\overline{)15.54}$

Number Correct

____
15

Math Test: Prime and Composite Numbers

Name ______________

Identify each number as either **prime (P)** or **composite (C)**.

1. 8 _____
2. 22 _____
3. 54 _____
4. 78 _____
5. 72 _____
6. 11 _____
7. 45 _____
8. 34 _____
9. 121 _____
10. 90 _____
11. 53 _____
12. 66 _____
13. 73 _____
14. 117 _____
15. 99 _____

Number Correct: /15

Math Test: Prime and Composite Numbers

Name ______________

Identify each number as either **prime (P)** or **composite (C)**.

1. 88 _____
2. 83 _____
3. 4 _____
4. 71 _____
5. 29 _____
6. 100 _____
7. 94 _____
8. 351 _____
9. 67 _____
10. 36 _____
11. 5 _____
12. 41 _____
13. 17 _____
14. 80 _____
15. 98 _____

Number Correct: /15

Name ______________

Math Test: Factors

Write all the factors for each given number below.

1. 14	
2. 45	
3. 88	
4. 36	
5. 24	
6. 76	
7. 60	
8. 96	
9. 44	
10. 11	
11. 50	
12. 33	
13. 55	
14. 10	
15. 66	

Number Correct: /15

Name ______________

Math Test: Greatest Common Factor

Find the greatest common factor of each set of numbers.

1. 8 and 48
2. 50 and 12
3. 21 and 42
4. 36 and 12
5. 4 and 36
6. 72 and 32
7. 14 and 63
8. 6 and 12
9. 5 and 25
10. 90 and 30
11. 35 and 70
12. 50 and 95
13. 20 and 40
14. 32 and 48
15. 7 and 70

Number Correct: /15

Name ______________

Math Test: Least Common Multiple

Find the lowest common multiple for each set of numbers below.

1.	8 and 20
2.	5 and 29
3.	4 and 7
4.	2, 5 and 10
5.	4 and 12
6.	6 and 17
7.	14 and 30
8.	6 and 12
9.	5 and 25
10.	8 and 12
11.	5 and 7
12.	3 and 8
13.	2 and 4
14.	6 and 11
15.	10 and 12

Number Correct: /15

Name ______________

Math Test: Percent of a Number

A. Find the percent of each number. Show your work.

1. 40% of 80	2. 15% of 140
3. 60% of 180	4. 50% of 90
5. 90% of 30	6. 75% of 160
7. 30% of 170	8. 60% of 20
9. 20% of 130	10. 30% of 70

Number Correct: /10

Name ______________

Math Test: Number Sense Grade 4

Circle the letter next to the correct answer.

1. Round the following number to the nearest 100. **675** A. 600 B. 700 C. 680	2. What is the least possible number using the following digits? **4961** A. 1469 B. 9641 C. 1946
3. What is this expanded number in standard form? **60 000 + 4000 + 800+ 50+ 2** A. 640 852 B. 60 452 C. 64 852	4. What is the numeral for the following number word? **three hundred sixty one** A. 316 B. 361 C. 3061
5. How many tens in 45? A. 5 tens B. 4 tens C. 3 tens	6. Find the number that is 1 more than 78. A. 80 B. 79 C. 77
7. Five hundred ninety is the number word for: A. 509 B. 519 C. 590	8. What number is 10 less than 875? A. 865 B. 870 C. 880
9. How many hundreds in 3095? A. 0 B. 3 C. 9	10. What is 7857 in expanded form? A. 7000 + 80 + 5 + 7 B. 7000 + 800 + 50 + 7 C. 7000 + 800 + 5 + 7

Name ______________

Math Test: Number Sense Grade 4

Circle the letter next to the correct answer.

11. Find the missing sign.

564 ◯ 645

A. <

B. =

C. >

12. Which is the least possible number using only the following digits?

4851

A. 1458

B. 8541

C. 1845

13. What is this expanded number in standard form?

10 000 + 4000 + 300 + 50

A. 140 350

B. 10 450

C. 14 350

14. What is the numeral for the following number word?

nine thousand one hundred thirteen

A. 9113

B. 9013

C. 913

15. Order these numbers from least to greatest.

2900, 450, 5621, 1178

A. 1178, 2900, 5621, 450

B. 450,1178, 2900, 5621

C. 5621, 2900,1178, 450

16. Find the missing sign.

5743 ◯ 2723

A. <

B. =

C. >

17. Find the missing sign.

760 ◯ 760

A. <

B. =

C. >

18. What number is 100 more than 892?

A. 900

B. 891

C. 992

19. How many tens in 8961?

A. 9

B. 1

C. 6

20. What is 4855 in expanded form?

A. 4000 + 80 + 5 + 5

B. 4000 + 800 + 50 + 5

C. 4000 + 800 + 5 + 5

Name ______________

Math Test: Number Sense Grade 4

Circle the letter next to the correct answer.

21. Find the missing sign.

6852 ◯ 4561

A. <
B. =
C. >

22. What is the greatest possible number using the following digits?

2915

A. 9512
B. 1259
C. 9521

23. What is this expanded number in standard form?

70 000 + 1000 + 500 + 20 + 4

A. 71 542
B. 71 524
C. 70 1524

24. What is the numeral for the following number word?

two thousand eighty seven

A. 2078
B. 287
C. 2087

25. Order these numbers from greatest to least.

2900, 450, 5621, 1178

A. 1178, 2900, 5621, 450
B. 1178, 450, 2900, 5621
C. 5621, 2900,1178, 450

26. Find the missing sign.

6905 ◯ 6800

A. <
B. =
C. >

27. Find the missing sign.

536 ◯ 1537

A. <
B. =
C. >

28. What number is 1000 more than 5745?

A. 6745
B. 6000
C. 5845

29. How many thousands in 5083?

A. 8
B. 3
C. 5

30. What is 4819 in expanded form?

A. 4000 + 80 + 1 + 9
B. 4000 + 800 + 1 + 9
C. 4000 + 800 + 10 + 9

Name ______________

Math Test: Number Sense Grade 5

Circle the letter next to the correct answer.

1. Find the missing sign.

16 622 ◯ 14 967

A. <

B. =

C. >

2. What is the greatest possible number using the following digits?

30149

A. 94 301

B. 93 410

C. 94 310

3. Which of the following groups of numbers is in order from largest to smallest?

A. 67, 10, 57, 87, 62

B. 77, 56, 30, 16, 4

C. 3, 56, 96, 89, 94

4. What is the numeral for the following number word?

twenty two thousand twenty

A. 2 220

B. 22 020

C. 22 222

5. What is this expanded number in standard form?

90 000 + 5000 + 700 + 80 + 1

A. 90 5781

B. 90 578

C. 95 781

6. Find the missing sign.

1845 ◯ 8000

A. <

B. =

C. >

7. Find the missing sign.

78 231 ◯ 78 231

A. <

B. =

C. >

8. What number is 10 000 more than 56 341?

A. 66 341

B. 56 441

C. 57 341

9. How many thousands in 34 783?

A. 8

B. 4

C. 5

10. What is 28 413 in expanded form?

A. 28 000 + 400 + 10 + 3

B. 2000 + 800 + 40 + 13

C. 20 000 + 8000 + 400 + 10 + 3

Name ____________

Math Test: Number Sense Grade 5

Circle the letter next to the correct answer.

11. What is the place value of the number in **bold?** 8,4**4**7

A. tens

B. ones

C. thousands

12. Which product is even?

A. 3 x 9

B. 4 x 10

C. 5 x 7

13. Which of the following group of numbers is in order from largest to smallest?

A. 87, 67, 62, 57, 10

B. 77, 26, 30, 16, 4

C. 3, 56, 96, 89, 94

14. What is the numeral for the following number word?

seventy two thousand fourteen

A. 7214

B. 72 014

C. 72 714

15. Which even number is between 40 and 60 and is a multiple of eight?

A. 41

B. 56

C. 64

16. Which two numbers are both factors of 48?

A. 2, 6

B. 3, 11

C. 14, 20

17. What is 704 rounded to the nearest ten?

A. 704

B. 700

C. 705

18. Which of the following numbers will have a remainder when it is divided by 11?

A. 66

B. 33

C. 97

19. Which number is prime?

A. 8

B. 33

C. 7

20. What is 18 411 in expanded form?

A. 18 000 + 400 + 10 + 1

B. 1000 + 800 + 40 + 1 + 1

C. 10 000 + 8000 + 400 + 10 + 1

Name ________________

Math Test: Number Sense Grade 6

Circle the letter next to the correct answer.

1. What is the greatest possible number using the following digits?

7 2 5 3 8

A. 82 753

B. 87 532

C. 23 578

2. Between what two numbers would 867 be found?

A. 863 and 903

B. 706 and 820

C. 790 and 860

3. What is the numeral for the following number word?

ninety seven thousand twelve

A. 971 200

B. 97 120

C. 97 012

4. Which of the following groups of numbers is in order from largest to smallest?

A. 1678, 1562, 1503, 1412, 157

B. 1562, 1678, 103, 1412, 157

C. 1575, 1412, 1503, 1678, 157

5. Find the missing sign.

386 457 ◯ 348 056

A. <

B. =

C. >

6. What is this expanded number in standard form?

90 000 + 5000 + 700 + 80 + 1

A. 90 5781

B. 90 578

C. 95 781

7. In which of the following numbers, does 5 have the greatest value?

A. 56 978

B. 74 506

C. 53

8. Find the missing sign.

145 210 ◯ 145 210

A. <

B. =

C. >

9. How many eggs in 6 ½ dozen?

A. 78 eggs

B. 72 eggs

C. 56 eggs

10. How many thousands in 734 783?

A. 8

B. 4

C. 5

Name ______________

Math Test: Number Sense Grade 6

Circle the letter next to the correct answer.

11. Which product is even? A. 7 x 8 B. 9 x 3 C. 5 x 7	12. What is the place value of the number in **bold?** 31 **4**57 A. hundreds B. ones C. thousands
13. How many of the following numbers are odd? 15, 76, 23, 91, 43 A. 3 B. 4 C. 5	14. How would you estimate 1572 + 6305 to the nearest hundred? A. 1600 + 6300 B. 1500 + 6200 C. 2000 + 6000
15. Which two numbers are both factors of 24? A. 2, 12 B. 3, 9 C. 3, 7	16. What does the 7 in the number 17 834 mean? A. 70 000 B. 7000 C. 700
17. What number is four thousand less than 227 901? A. 231 901 B. 227 501 C. 223 901	18. What is 4653 rounded to the nearest thousands? A. 4600 B. 4000 C. 5000
19. What is 847 915 in expanded form? A. 800 000 + 47 000 + 915 B. 800 000 + 40 000 + 7000 + 900 + 1 + 5 C. 800 000 + 40 000 + 7000 + 900 + 10 +5	20. Which number is prime? A. 342 B. 17 C. 45

Name ______________

Math Test: Fractions

A. Fill in the missing number to complete each equivalent fraction.

1. $\frac{1}{4} = \frac{\_\_\_}{28}$	2. $\frac{7}{12} = \frac{\_\_\_}{48}$	3. $\frac{3}{8} = \frac{\_\_\_}{24}$
4. $\frac{7}{10} = \frac{\_\_\_}{60}$	5. $\frac{2}{4} = \frac{\_\_\_}{16}$	6. $\frac{7}{9} = \frac{\_\_\_}{72}$
7. $\frac{5}{6} = \frac{\_\_\_}{24}$	8. $\frac{\_\_\_}{5} = \frac{16}{20}$	9. $\frac{3}{10} = \frac{\_\_\_}{20}$
10. $\frac{12}{16} = \frac{\_\_\_}{32}$	11. $\frac{\_\_\_}{4} = \frac{18}{36}$	12. $\frac{24}{36} = \frac{\_\_\_}{6}$
13. $\frac{12}{16} = \frac{\_\_\_}{8}$	14. $\frac{15}{20} = \frac{3}{\quad}$	15. $\frac{18}{21} = \frac{\_\_\_}{7}$

Number Correct: /15

Name ______________

Math Test: Fractions

A. Write an equivalent fraction for each of the following.

1. $\frac{3}{8} =$	2. $\frac{1}{4} =$	3. $\frac{2}{3} =$

B. Change each mixed number to an improper fraction.

1. $4\frac{1}{5} =$	2. $2\frac{3}{7} =$	3. $3\frac{2}{3} =$

C. Write each fraction in its simplest form.

1. $\frac{6}{9} =$	2. $\frac{4}{12} =$	3. $\frac{30}{90} =$
4. $\frac{25}{35} =$	5. $\frac{10}{20} =$	6. $\frac{9}{72} =$

D. Arrange these fractions in order from <u>least</u> to <u>greatest</u>.

$\frac{2}{8}$, $\frac{1}{8}$, $\frac{5}{8}$

E. There were 20 flowers in a vase. 12 of the flowers were tulips. The rest of the flowers were roses. What fraction of the flowers were roses? Write your answer in simplest form.

Name ______________

Math Test: Decimals

A. Write each as a decimal.

1. $\frac{4}{10}$ _____
2. $\frac{35}{100}$ _____
3. $\frac{2}{10}$ _____
4. $\frac{67}{100}$ _____

5. one tenth ________________________________

6. twelve hundredths ________________________________

7. eighty-nine hundredths ________________________________

8. three tenths ________________________________

B. Compare using < > or =

9. 4.03 ______ 4.10

10. 6.5 _____ 6.56

11. 0.6 ______ 1.3

12. 1.97 _____ 1.97

13. 9.5 ______ 9.05

14. 0.8 _____ 0.63

15. 8.095 ______ 8.0

16. 2.5 _____ 2.05

C. Order the decimals from <u>greatest</u> to <u>least</u>.

17. 8.77, 7.1, 0.99, 3.1, 8.12 ________________________________

18. 3.9, 1.80, 2.5, 0.2, 0.78 ________________________________

D. Order the decimals from <u>least</u> to <u>greatest</u>.

19. 9.302, 9.6, 9.51, 8.01 ________________________________

20. 2.7, 9.71, 8.8 ________________________________

Name ______________

Math Test: Fractions

A. Add the following fractions and write the answer in simplest form.

1. $\frac{2}{4} + \frac{1}{4} =$

2. $\frac{1}{5} + \frac{6}{5} =$

3. $\frac{1}{2} + \frac{2}{6} =$

4. $\frac{1}{6} + \frac{2}{5} =$

B. Subtract the following fractions and write the answer in simplest form.

1. $\frac{4}{6} - \frac{1}{6} =$

2. $\frac{8}{10} - \frac{3}{10} =$

3. $\frac{2}{2} - \frac{5}{14} =$

4. $\frac{4}{7} - \frac{1}{2} =$

C. Multiply the following fractions.

1. $4 \times \frac{2}{3} =$

2. $\frac{2}{4} \times 3 =$

3. $\frac{1}{16} \times \frac{2}{4} =$

4. $\frac{1}{9} \times \frac{4}{3} =$

D. Divide the following fractions.

1. $6 \div \frac{1}{2} =$

2. $2 \div \frac{5}{7} =$

3. $\frac{1}{3} \div \frac{2}{4} =$

4. $\frac{2}{4} \div \frac{1}{6} =$

Name ______________

Math Test: Fractions, Percents, Decimals and Ratios

A. Complete the chart:

Fraction	Denominator of 100	Percentage	Decimal	Ratio
1. $\frac{8}{20}$				
2. $\frac{7}{25}$				
3. $\frac{16}{10}$				
4. $\frac{3}{5}$				

B. Complete the following by looking at the shaded squares:

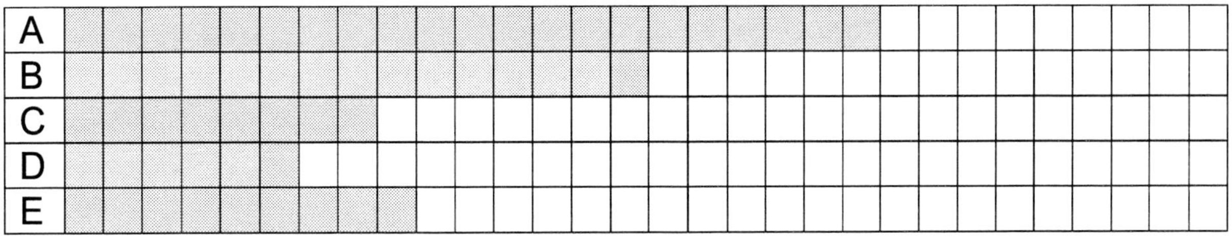

A																														
B																														
C																														
D																														
E																														

(Note: There are 30 spaces in each row)

1. Which row is 0.30 shaded? ______________________________
2. Which row is 20% shaded? ______________________________
3. What percentage of row B is shaded? ______________________
4. Which row is between ¼ and ½ shaded? __________________
5. Which row is 0.70 shaded? ______________________________
6. What ratio of row C is shaded? ___________________________

Name ______________

Math Test: How Much Money?

A. Count and write the value of each amount of money

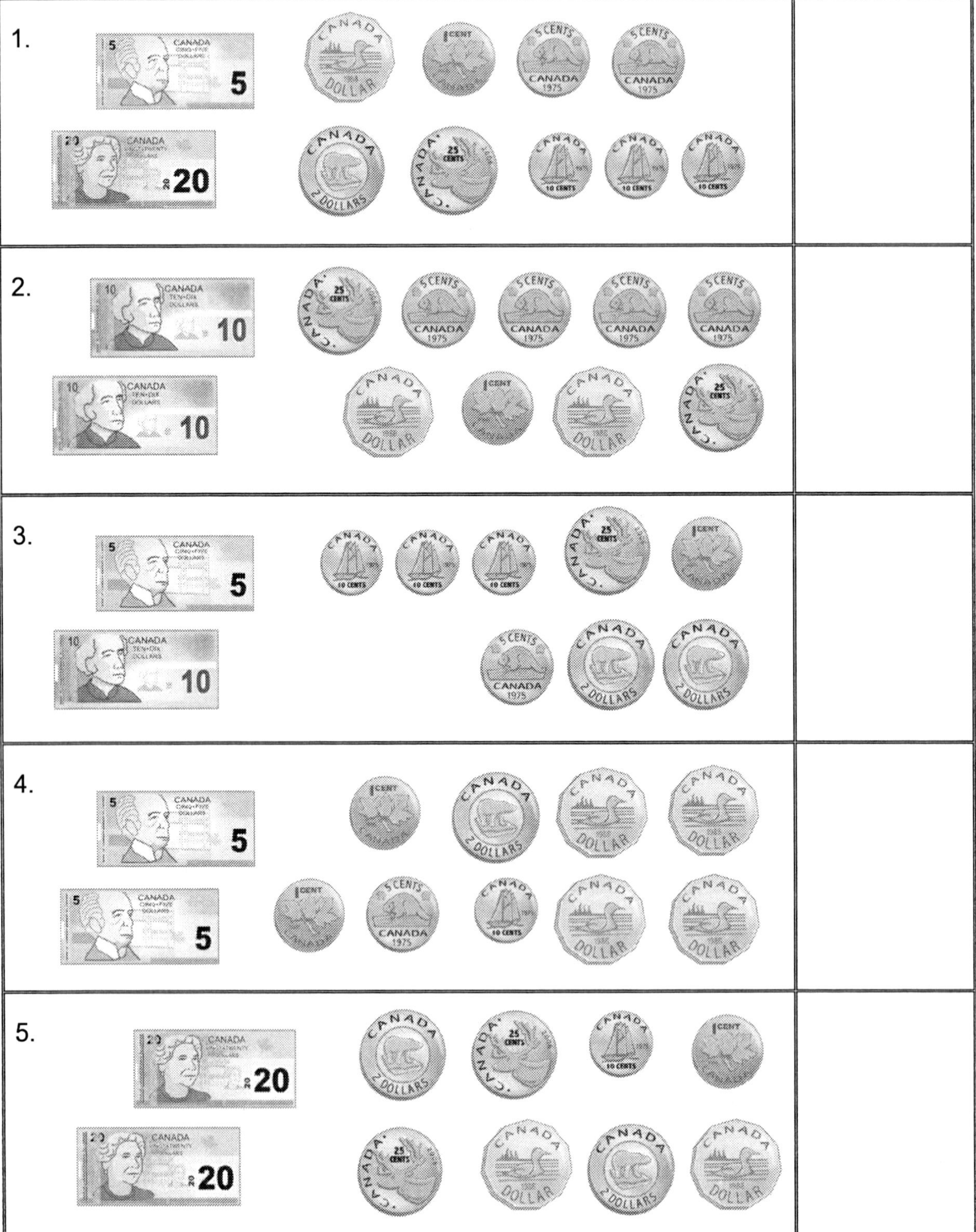

1.	
2.	
3.	
4.	
5.	

Name ________________

Math Test: How Much Money?

A. Count and write the value of each amount of money

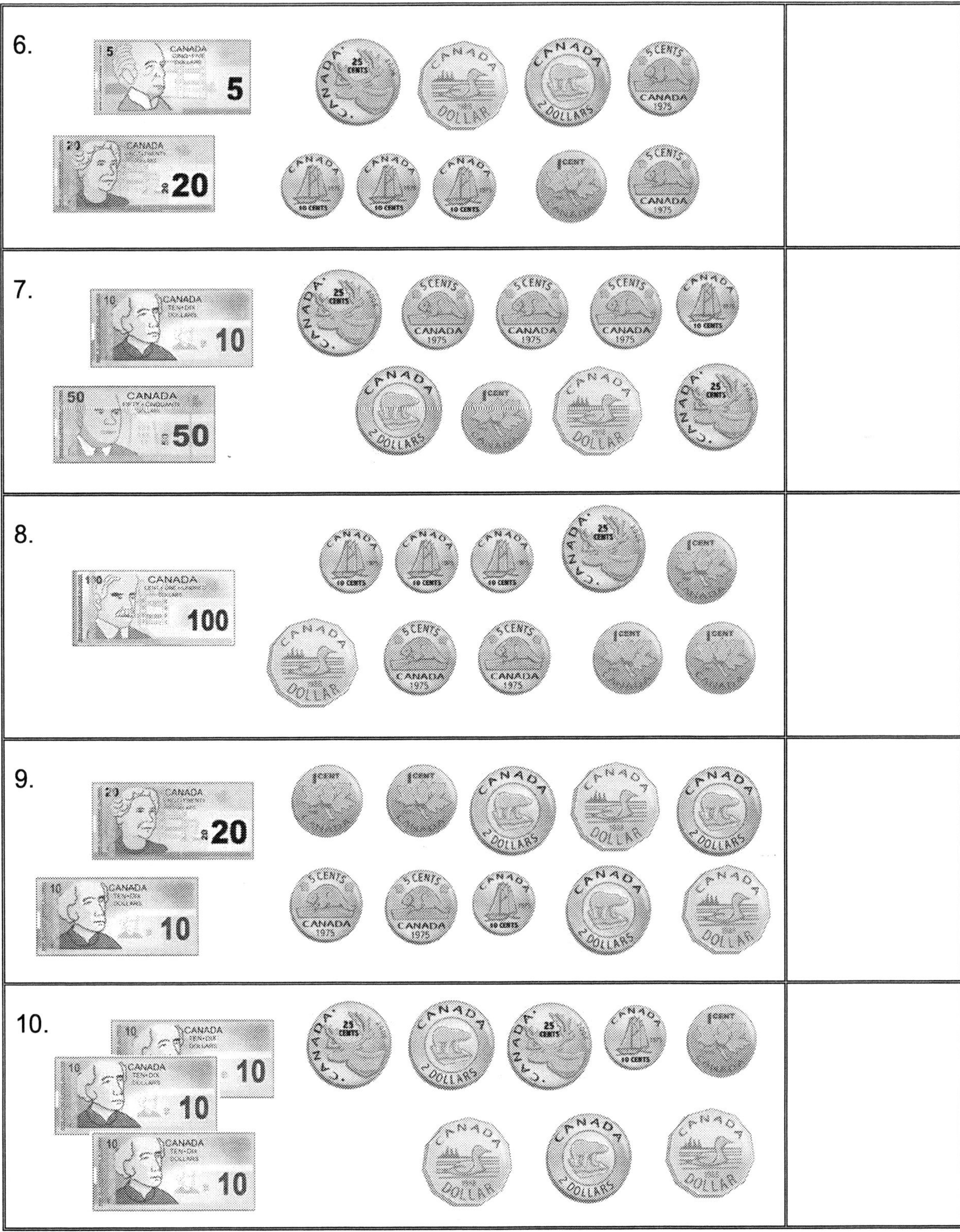

Number Correct: /15

Name ____________

Math Test: Money

A. Solve the following. Show your thinking.

1. How many dimes are in 10 toonies?

2. Katelyn bought a pair of skates for $74.22. She paid the cashier 6 bills and 4 coins. What were the bills and coins Katelyn used?

3. How much money in total?

   7 twenty-dollar bills, 1 toonie, 5 loonies, 3 quarters, 7 dimes, and 9 nickels.

4. Madelyn bought 4 bunches of tulips. Each bunch cost $3.20. She paid with $20.00. What was Madelyn's change?

5. Megan bought 3 T-shirts. Each T-shirt cost $19.60. How much did the T-shirts cost altogether?

Patterning

And

Algebra

Tests

Name ______________

Math Test: Order of Operations

A. Simplify.

1. 23 × 12 + 66 - 5	2. 84 ÷ 6 + 2 × 5
3. 81 - 3 × 5	4. 72 ÷ 4 + 4 + 23
5. 6 × 49 + 26 × 31	6. 94 ÷ 2 - 1
7. 81 ÷ 3 × 3 + 1	8. 9 × 3 + 1 + 10
9. 24 ÷ 8 + 1	10. 67 × 5 - 3 - 1

Name _______________

Math Test: Order of Operations

A. Simplify.

11. $7 \times 50 + (32 \div 8) - 30$	12. $88 \div 2 + 2 + 14$
13. $(77 \div 7) \times (12 \div 4) - 3 \times 5$	14. $9 \times 42 + (1 \times 5)$
15. $(5 \times 29) + (31 - 5) + 15$	16. $75 - 4 - (36 \div 2) + 12$
17. $(5 \times 3) + (45 \div 9) - 3$	18. $12 \times 3 + 8 + 6$
19. $(5 + 51) - (4 + 38) + 6$	20. $(56 \times 33) + (27 - 1 \times 4)$

Number Correct: /20

Name ______________

Math Test: Patterning and Algebra Grade 4

Circle the letter next to the correct answer.

1. What is next in the pattern?

11, 21, 31, ___, ___

A. 40, 51
B. 32, 35
C. 41, 51

2. Find the missing number.

222, 224, 226, _____, 230

A. 231
B. 232
C. 228

3. Which number sentence has the same product as 6 x 4?

A. 9 x 3
B. 10 x 2
C. 8 x 3

4. What are the next two creatures in the pattern?

A. 
B. 
C.

5. What number comes next?

2, 4, 6, 8, 10, ____

A. 12
B. 11
C. 14

6. What is the pattern rule?

10, 20, 40, 80

A. double each time
B. multiply by 3
C. add 20

7. What is the next number, if the pattern rule is multiply by 5?

7, ____

A. 35
B. 12
C. 30

8. Each box holds 9 cookies. How many cookies in 7 boxes?

A. 64 cookies
B. 63 cookies
C. 72 cookies

9. What is the missing number?

_____ ÷ 7 = 7

A. 54
B. 49
C. 63

10. Which number sentence is the same as 81 ÷ 9?

A. 32 ÷ 4
B. 48 ÷ 8
C. 27 ÷ 3

Name ______________

Math Test: Patterning and Algebra Grade 4

Circle the letter next to the correct answer.

11. What is next in the pattern?

15, 20, 25, ___, ___

A. 50, 75
B. 30, 35
C. 35, 45

12. Find the missing number.

674, 684, 694, __

A. 704
B. 699
C. 714

13. What should replace the _____ to make the following sentence true?

13 + 86 = 9 _____ 11?

A. x
B. +
C. -

14. Predict what the 19th figure will be in this pattern.

A.
B.
C.

15. What is another way to write

7 + 7 + 7 + 7

A. 7 ÷ 7
B. 7 x 7
C. 4 x 7

16. What is the pattern rule?

30, 60, 90, 120

A. double each time
B. multiply by 3
C. add 30

17. What is the next number if the pattern rule is subtract 20?

140, ____

A. 130
B. 120
C. 150

18. John runs 6 km every day. How many km will John run in 23 days?

A. 138 km
B. 120 km
C. 29 km

19. What is the missing number?

_____ ÷ 11 = 55

A. 605
B. 610
C. 615

20. Which number sentence has the same answer as 49 ÷ 7 =

A. 32 ÷ 4
B. 48 ÷ 8
C. 21 ÷ 3

Name ____________

Math Test: Patterning and Algebra Grade 5

Circle the letter next to the correct answer.

1. What is next in the pattern?

610, 710, 810 , ____, ____

A. 910, 1001
B. 910,1010
C. 910, 1000

2. What kind of pattern is this?

1 2 1 2 1 2 1 2 1 2 1 2 1 2 1 2

A. repeating
B. shrinking
C. growing

3. There were 342 people who attended a movie. If each ticket cost $6.00, how much was collected?

A. $2050
B. $2252
C. $2052

4. Predict what the 43rd animal will be in this pattern.

A. 
B.
C.

5. What number comes next?

484, 474, 464, _____

A. 474
B. 454
C. 444

6. What is the pattern rule?

10, 30, 90, 270

A. double each time
B. multiply by 3
C. add 20

7. What is the next number, if the pattern rule is multiply 9?

11, _____

A. 90
B. 19
C. 99

8. What should replace the _____ to make the following equation true?

8 × 11 = 97 _____ 9

A. +
B. -
C. ÷

9. What is the missing number?

11 X _____ = 66

A. 16
B. 6
C. 7

10. In which number sentence does a 8 make the equation true?

A. 16 x ____ = 32
B. 16 ÷ ____ = 2
C. 24 + ____ = 30

Name _______________

Math Test: Patterning and Algebra Grade 5

Circle the letter next to the correct answer.

11. What is next in the pattern?

0, 25, ___, ___, 100

A. 50, 75

B. 30, 35

C. 35, 45

12. What number is missing from the following sequence?

18, 29, 40, 51, ____, 73

A. 69

B. 72

C. 62

13. What should replace the _____ to make the following equation true?

43 + 37 = 8 _____ 10

A. x

B. +

C. -

14. Predict what the 23rd figure will be in this pattern.

A.

B.

C.

15. Each spider has 8 legs. How many legs do 15 spiders have?

A. 96 legs

B. 120 legs

C. 128 legs

16. What is the pattern rule?

20, 40, 80, 160

A. double each time

B. multiply by 3

C. add 30

17. What is the next number if the pattern rule is subtracted by 80?

140, ____

A. 40

B. 80

C. 60

18. Dianne runs 8 km every day. How many km will Dianne run in 23 days?

A. 184 km

B. 120 km

C. 29 km

19. What is the missing number?

_____ ÷ 9 = 8

A. 72

B. 78

C. 111

20. Which number sentence has the same product as 64 ÷ 8

A. 32 ÷ 4

B. 48 ÷ 8

C. 27 ÷ 3

Name ______________

Math Test: Patterning and Algebra Grade 6

Circle the letter next to the correct answer.

1. What is next in the pattern?

1010, 910, 810 , ___, ___

A. 710, 610

B. 710, 601

C. 700, 600

2. What kind of pattern is this?

1 2 3 4 5 6 7 8 9 1 0 1 1

A. repeating

B. shrinking

C. growing

3. There were 556 people who attended a hockey game. If each ticket cost $8.00 how much money was collected?

A. $4440

B. $4448

C. $4548

4. Predict what the 23rd animal will be in this pattern.

A.

B.

C.

5. What number comes next?

474, 464, 454, ____

A. 474

B. 454

C. 444

6. What is the pattern rule?

80, 60, 40, 20

A. double each time

B. multiply by 3

C. subtract by 20

7. What is the next number if the pattern rule is add 8?

11, ____

A. 90

B. 19

C. 99

8. What should replace the _____ to make the following equation true?

8 + 3 = 99 _____ 9

A. +

B. -

C. ÷

9. What is the missing number?

11 X ____ = 176

A. 16

B. 6

C. 7

10. In which number sentence does an 8 make the equation true?

A. 4 x ___ = 32

B. 16 ÷ ___ = 4

C. 24 + ___ = 30

Name ______________

Math Test: Patterning and Algebra Grade 6

Circle the letter next to the correct answer.

11. What is next in the pattern?

45, 35, 25, ___, ___

A. 50, 75

B. 30, 35

C. 15, 5

12. What should replace the _____ to make the following equation true?

20 + 36 = 7 _____ 8

A. x

B. +

C. ÷

13. What number is missing from the following sequence?

1, 36, 1296, _____, 1 679 616

A. 2592

B. 46 656

C. 40 888

14. Predict what the 20th figure will be in this pattern.

A.

B.

C.

15. Rob has 25 bags of jelly beans. Each bag has 125 jelly beans. Which number sentence finds the total jelly beans Rob has?

A. 25 x 125

B. 25 + 125

C. 125 ÷ 25

16. What is the pattern rule?

270, 90, 30, 10

A. double each time

B. divide by 3

C. add 30

17. What is the next number, if the pattern rule is divide by 12?

144, ____

A. 12

B. 120

C. 122

18. Indicate the value of the letter *C*.

$$4 \times C - 4 = 16$$

A. $C = 6$

B. $C = 4$

C. $C = 5$

19. There are 12 roses in each vase. How many roses in 23 vases?

A. 256

B. 250

C. 276

20. Which number sentence is the same as the product of 13 x 5?

A. 60 - 6

B. 22 + 30

C. 100 – 35

Geometry Tests

Name ________________

Math Test: Naming Shapes

A. Name each shape.

1. 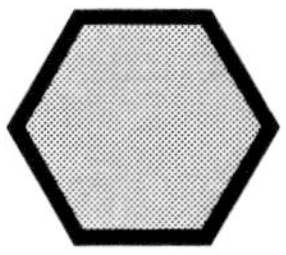 ________________	2. 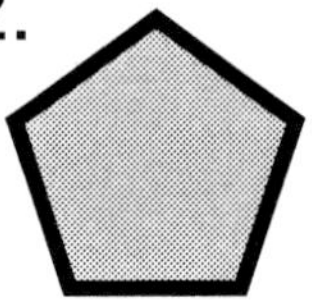 ________________
3.  ________________	4.  ________________
5. 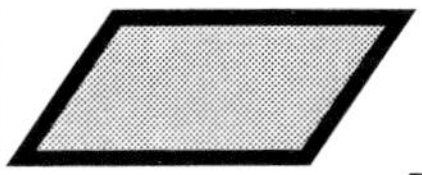 ________________	6. 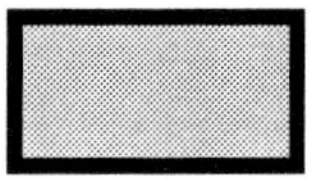 ________________
7.  ________________	8.  ________________
9.  ________________	10. 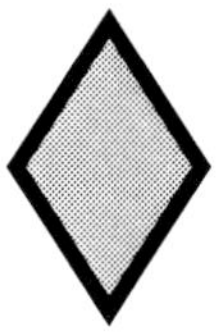 ________________

rectangle	**square**	**pentagon**	**circle**	**hexagon**
octagon	**triangle**	**rhombus**	**parallelogram**	**trapezoid**

Name ________________

Math Test: Classifying and Sorting Shapes

Read the sorting rule.
Mark the shapes that follow each rule.

1. Shapes with more than 4 sides.

  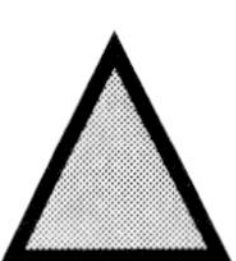  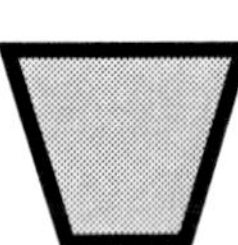

2. Shapes with less than 4 sides.

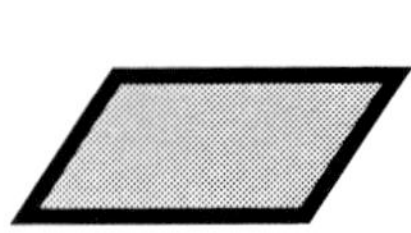   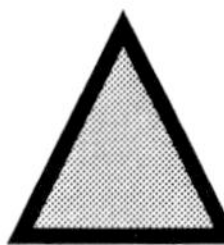

3. Shapes with more than 4 vertices.

 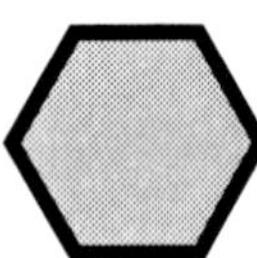  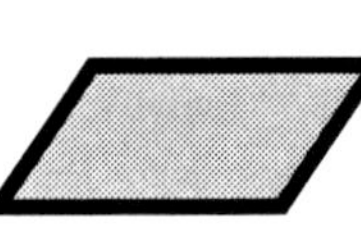 

4. Shapes with parallel lines.

 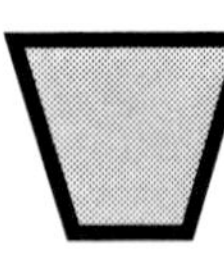 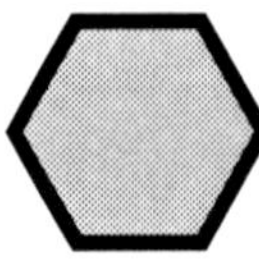  

5. Shapes that are polygons.

   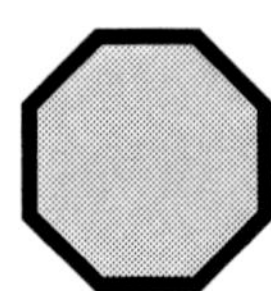 

Name ______________

Math Test: Classifying and Sorting Shapes

Read the sorting rule.
Mark the shapes that follow each rule.

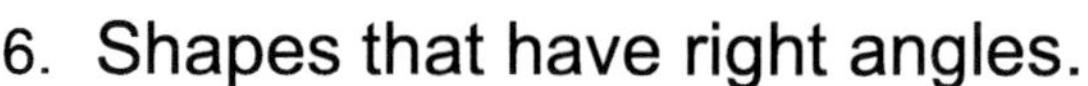

6. Shapes that have right angles.

7. Shapes that are not polygons.

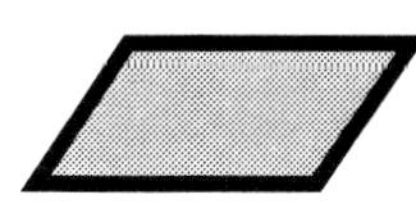 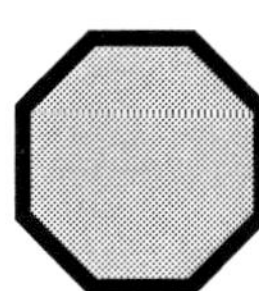  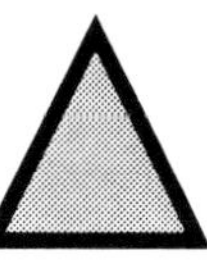 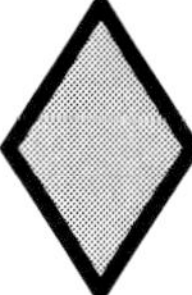

8. Shapes with more than 5 vertices.

 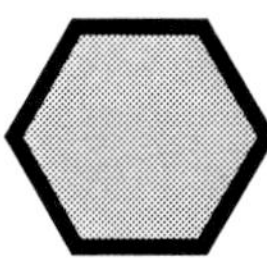  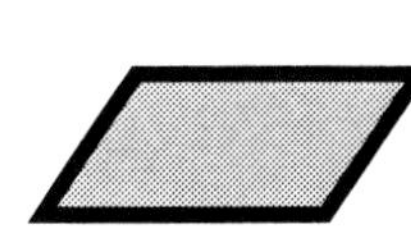 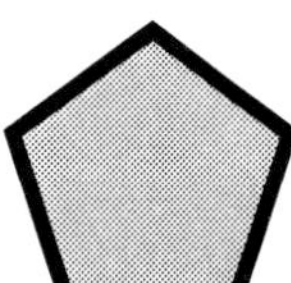

9. Shapes that have equal sides.

 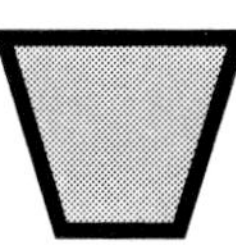 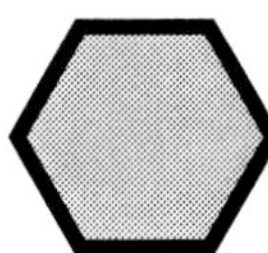  

10. Shapes that are quadrilaterals.

   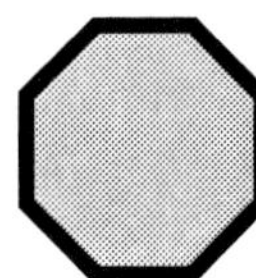 

Name ______________

Math Test: 3D Figures

Draw a line to match each 3D figure to an object that matches.

1. 	A. 
2. 	B. 
3. 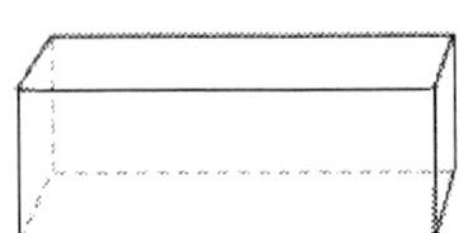	C. 
4. 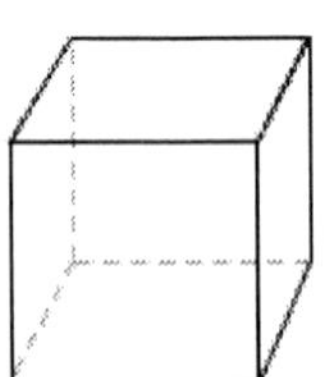	D. 
5. 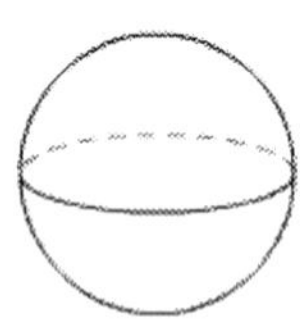	E. 
6. 	F. 

Name ______________

Math Test: 3D Figures

A. Complete.

1. This is a ___________________________.

How many? faces _______ edges______ vertices ______

2. This is a ___________________________.

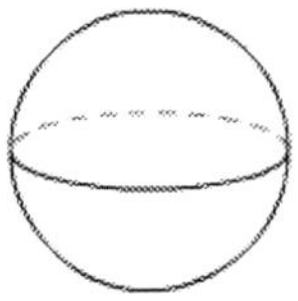

How many? faces _______ edges______ vertices ______

3. This is a ___________________________.

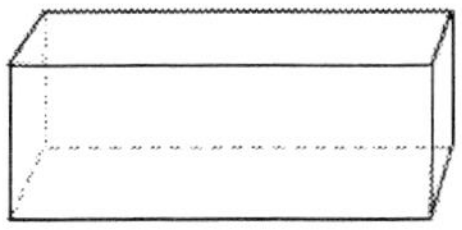

How many? faces _______ edges______ vertices ______

4. This is a ___________________________.

How many? faces _______ edges______ vertices ______

5. This is a ___________________________.

How many? faces _______ edges______ vertices ______

6. This is a ___________________________.

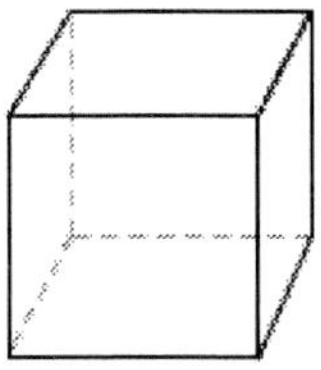

How many? faces _______ edges______ vertices ______

Name ______________

Math Test: Symmetry

A. Show and write the lines of symmetry for the following letters.

1. H How many? _____	2. E How many? _____	3. Q How many? _____
4. N How many? _____	5. Y How many? _____	6. B How many? _____
7. A How many? _____	8. R How many? _____	9. T How many? _____
10. G How many? _____	11. W How many? _____	12. V How many? _____
13. J How many? _____	14. C How many? _____	15. P How many? _____

Number Correct: /15

Name ______________

Math Test: Angles

A. Complete.

1. An angle that is more than 90° is called ____________________.

2. An angle that is 90° is called ______________________________.

3. An angle that is less than 90° is called ______________________.

B. Classify each of the following angles: acute, obtuse or right.

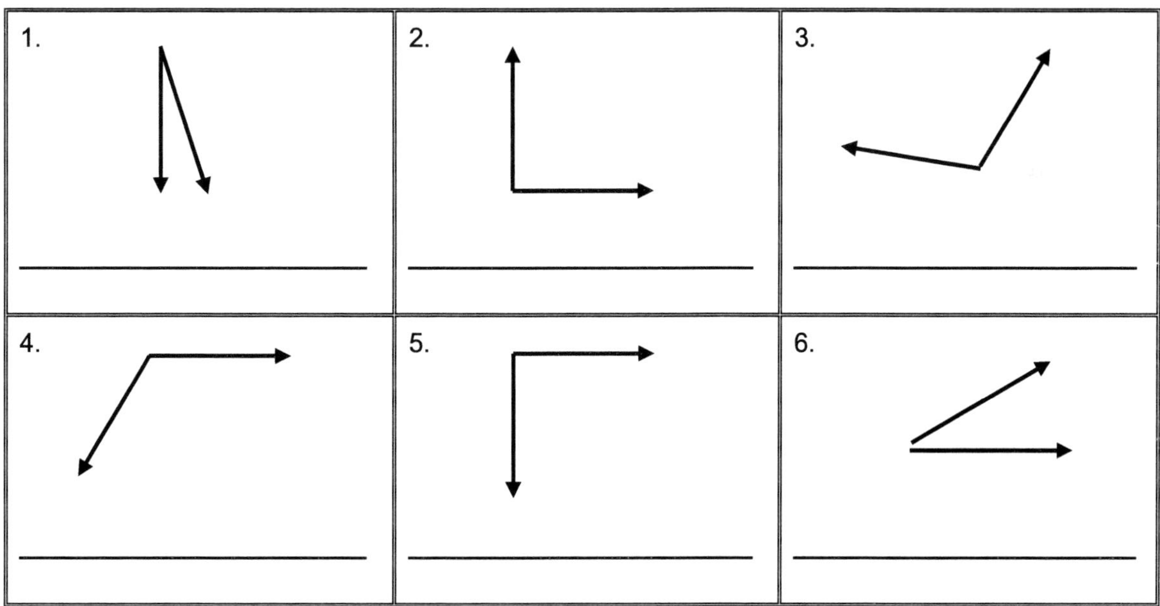

C. What is the size in degrees of the remaining angle if a triangle has angles of 60° and 90°?

____

D. What is the size in degrees of the remaining angle if a parallelogram has angles of 120°, 60°, and 60°?

____

E. Which angle is 90 degrees?

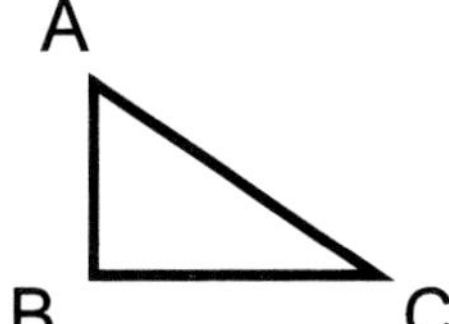

____

Name ______________

Math Test: Classifying Angles

A. Classify the angles as acute, obtuse, straight, or right.

1. 30 ° angle ______________	2.  ______________
3. 180 ° angle ______________	4. 20 ° angle ______________
5. 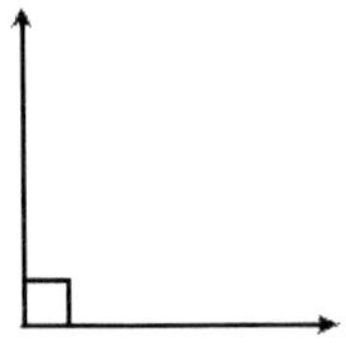 ______________	6. 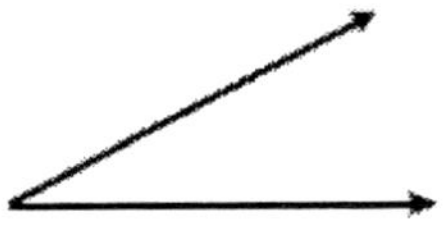 ______________
7. 90 ° angle ______________	8. 105 ° angle ______________
9. 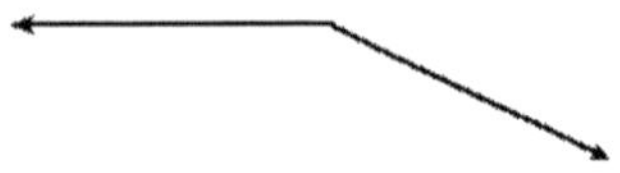 ______________	10.  ______________

Number Correct: /15

Name ______________

Math Test: Angles

A. Calculate the measure of the missing angle and name the triangle.

1.

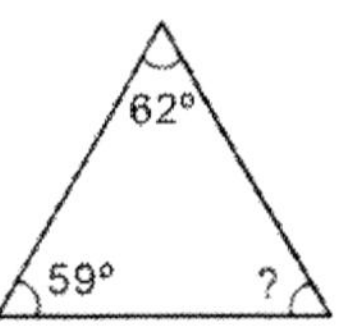

measure of angle ______________

type of triangle ________________

2.

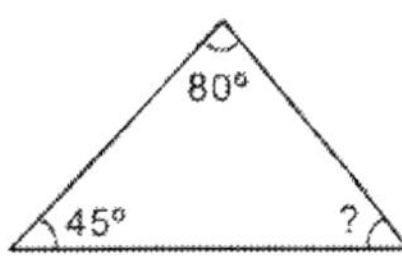

measure of angle ______________

type of triangle ________________

3.

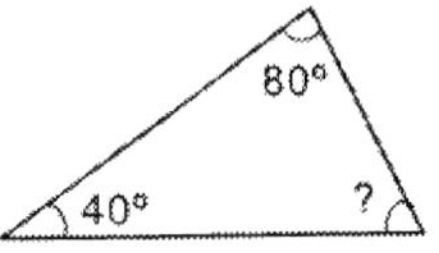

measure of angle ______________

type of triangle ________________

4.

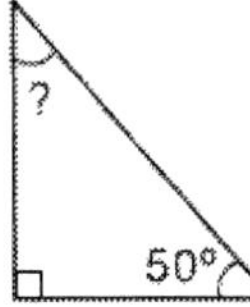

measure of angle ______________

type of triangle ________________

5.

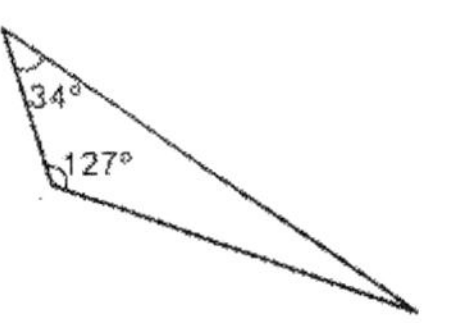

measure of angle ______________

type of triangle ________________

6.

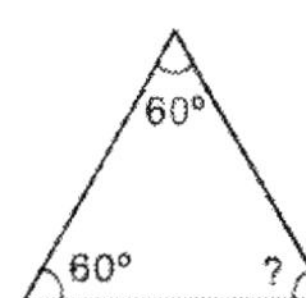

measure of angle ______________

type of triangle ________________

7.

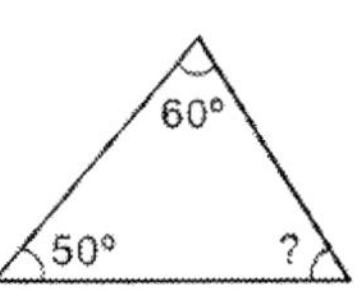

measure of angle ______________

type of triangle ________________

8.

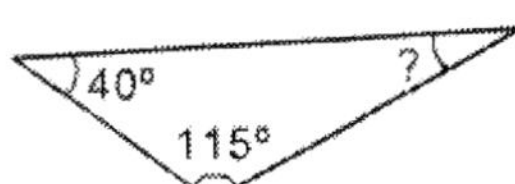

measure of angle ______________

type of triangle ________________

9.

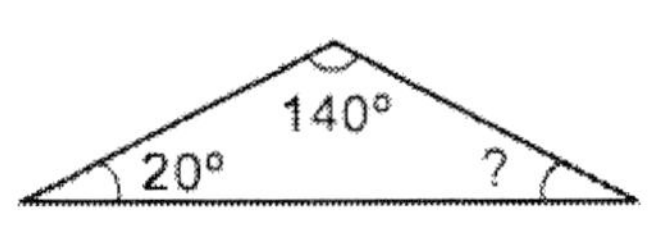

measure of angle ______________

type of triangle ________________

10.

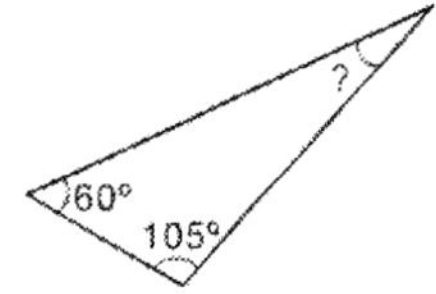

measure of angle ______________

type of triangle ________________

Name ________________

Math Test: Geometry Grade 4

Circle the letter next to the correct answer.

1. What is the name of this polygon?

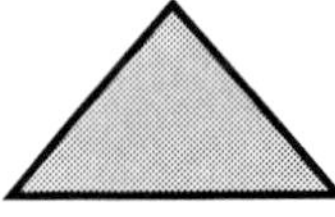

A. triangle B. circle C. rhombus

2. Which figure shows a line of symmetry?

A. 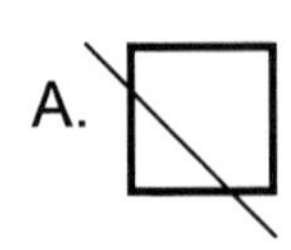 B. 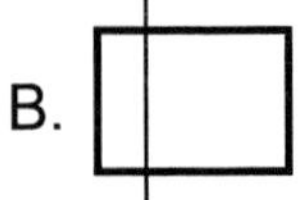 C. 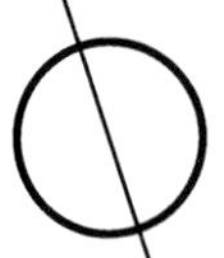

3. What is the name of this 3D figure?

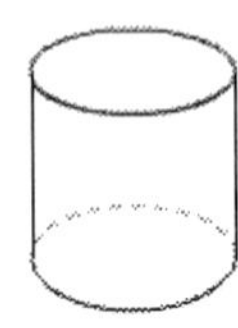

A. cylinder B. sphere C. pyramid

4. What is the name of this shape?

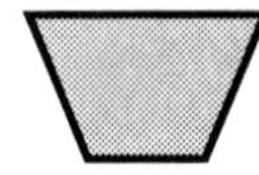

A. pentagon B. hexagon C. trapezoid

5. Which 3D figure has no edges?

A. 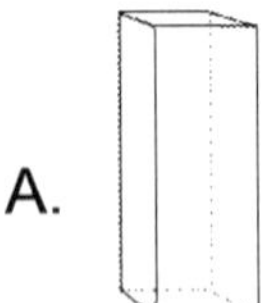 B. 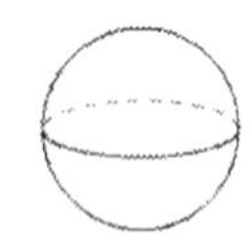 C. 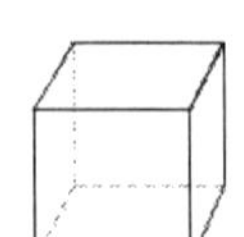

6. Look at the shapes. Choose flip, slide or turn.

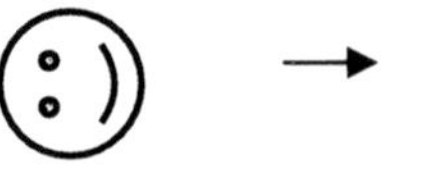

A. flip B. slide C. turn

7. How many lines of symmetry does this letter have?

M

A. 1 B. 2 C. 3

8. What shape is a quadrilateral?

A. octagon B. triangle C. square

9. Which of these are parallelograms?

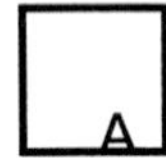

B

A. shapes B and A B. shape D C. shape B

10. Which 3D figure is a cube?

A. 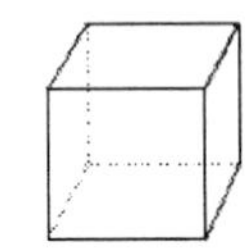 B.  C.

Name ________________

Math Test: Geometry Grade 4

Circle the letter next to the correct answer.

11. An angle of 140° is called?

A. obtuse B. right C. acute

12. A quadrilateral is a polygon with how many sides?

A. 5 sides B. 6 sides C. 4 sides

13. What is the name of the polygon with five sides?

A. hexagon B. pentagon C. circle

14. What is the name of this 3D figure

A. cube B. sphere C. pyramid

15. Look at the shapes. Choose flip, slide or turn.

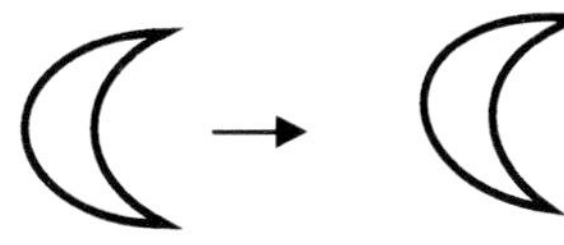

A. flip B. slide C. turn

16. Which pair of shapes look congruent?

W

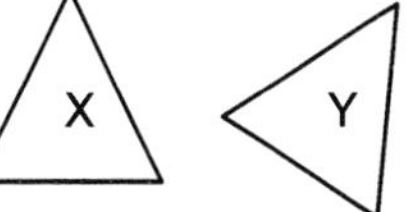

Z

A. W and X B. X and Y C. W and Z

17. How many edges?

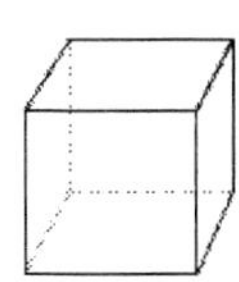

A. 10 B. 11 C. 12

18. How many of the following pairs of line intersect?

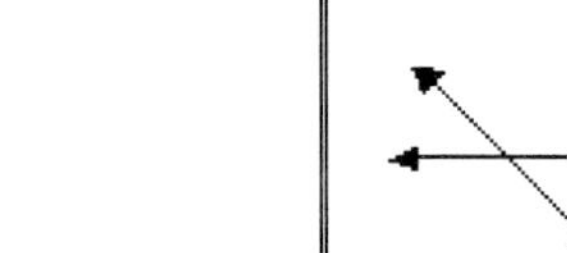

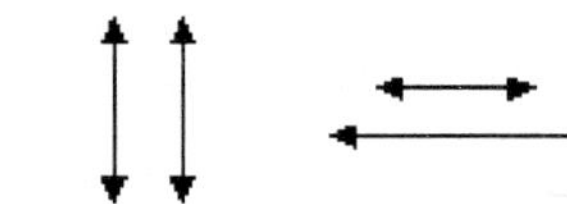

A. 1 B. 2 C. 3

19. What 3D figure does this object look like?

A. cone B. sphere C. cylinder

20. What 3D figure could be made from these pieces?

A. cylinder B. cone C. pyramid

Name ______________

Math Test: Geometry Grade 5

Circle the letter next to the correct answer.

1. How many vertices does a triangle have?

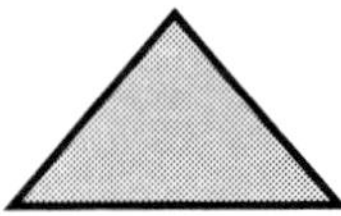

A. 3 B. 6 C. 2

2. Which figure shows a line of symmetry?

A. 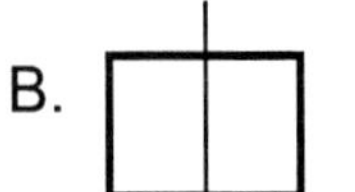 B. C. 

3. What is the name of this 3D figure?

A. cube B. sphere C. pyramid

4. Classify the following pair of lines.

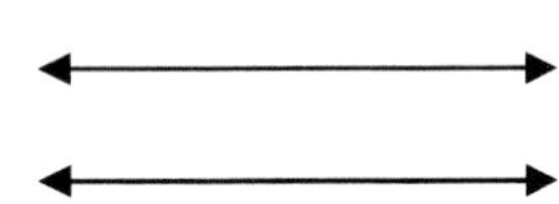

A. intersecting B. parallel C. perpendicular

5. Which 3D figure has 8 edges?

A.  B. C.

6. Look at the shapes. Choose flip, slide or turn.

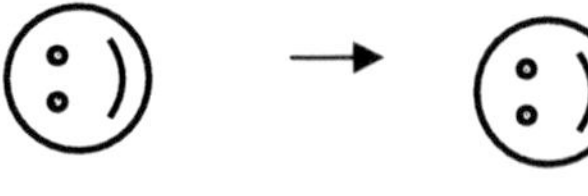

A. flip B. slide C . turn

7. How many lines of symmetry does this letter have?

A. 1 B. 2 C. 3

8. Which pair of figures is congruent?

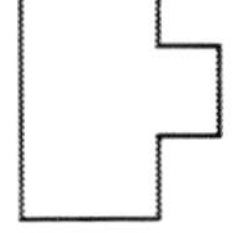

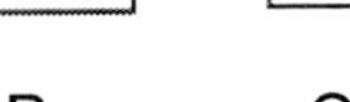

P Q R

A. P and Q B. P and R C. Q and P

9. Which of these is a trapezoid?

D

A. shape B B. shape D C. shape B

10. Which 3D figure is a pyramid?

A. 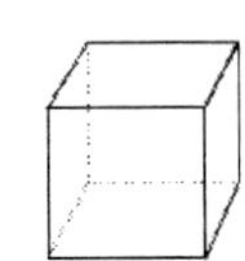 B.  C.

Name ________________

Math Test: Geometry Grade 5

Circle the letter next to the correct answer.

11. An angle of 120° is called?

A. obtuse B. right C. acute

12. A rhombus is a polygon with how many sides?

A. 5 sides B. 6 sides C. 4 sides

13. Which shape has 2 pairs of parallel sides, and 2 pairs of angles that are equal?

A.  B. 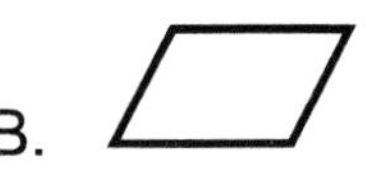 C.

14. Classify the following triangle.

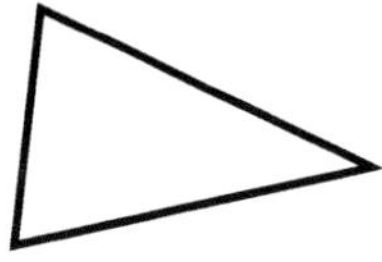

A. scalene B. isosceles C. equilateral

15. Look at the shapes. Choose flip, slide or turn.

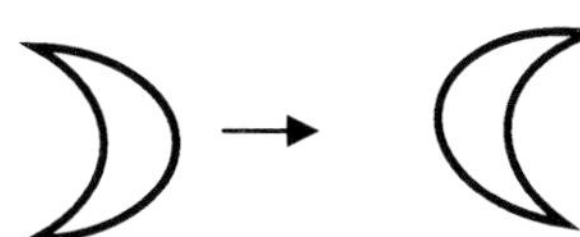

A. flip B. slide C. turn

16. Which pair of shapes look congruent?

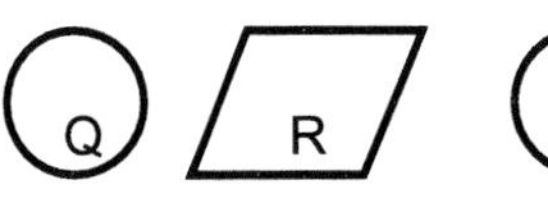

A. R and T B. Q and S C. Q and T

17. How many faces?

A. 5 B. 4 C. 12

18. How many of the following pairs of line are parallel?

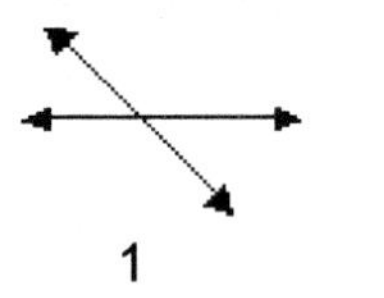

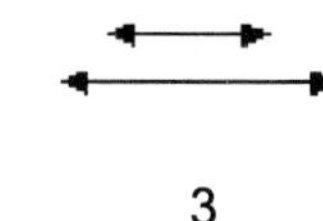

A. 1 B. 2 C. 3

19. What 3D figure does this object look like?

A. cube B. sphere C. cylinder

20. What 3D figure could be made from these pieces?

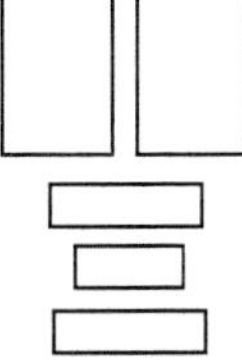

A. cylinder B. rectangular prism C. pyramid prism

Name ______________

Math Test: Geometry Grade 6

Circle the letter next to the correct answer.

1. Which figure shows a line of symmetry?

A. B. C.

2. What is the name of this polygon?

A. hexagon B. circle C. rhombus

3. Classify the following group of lines.

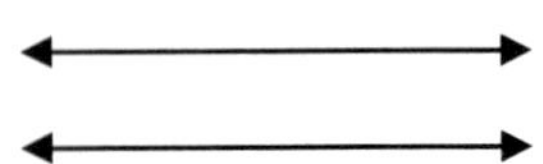

A. intersecting B. parallel C. perpendicular

4. What is the name of this 3D figure?

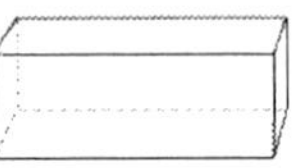

A. rectangular prism B. sphere C. pyramid

5. Look at the shapes. Choose flip, slide or turn.

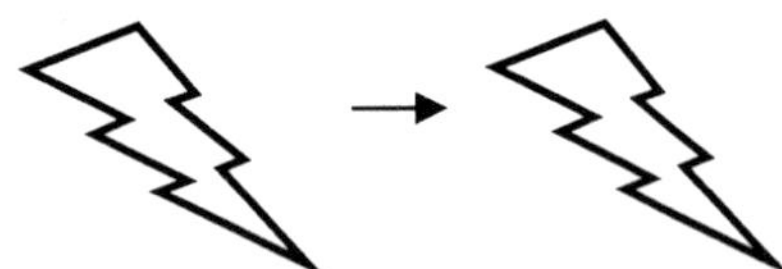

A. flip B. slide C. turn

6. Which 3D figure has 12 edges?

A.  B. C. 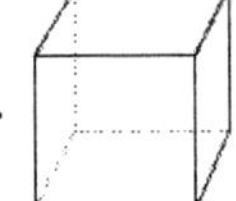

7. What shape is **<u>not</u>** a quadrilateral?

A. rectangle B. rhombus C. circle

8. How many lines of symmetry does this letter have?

A. 1 B. 2 C. 3

9. Which 3D figure is has no edges?

A.  B. 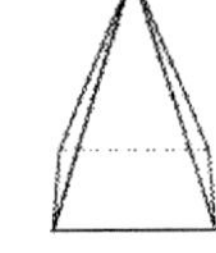 C. 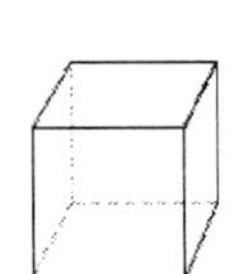

10. What 3D figure could be made from these pieces?

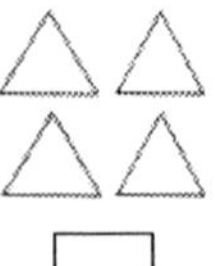

A. cylinder B. rectangular prism C. pyramid

Name ________________

Math Test: Geometry Grade 6

Circle the letter next to the correct answer.

11. What is the measure of the missing angle?

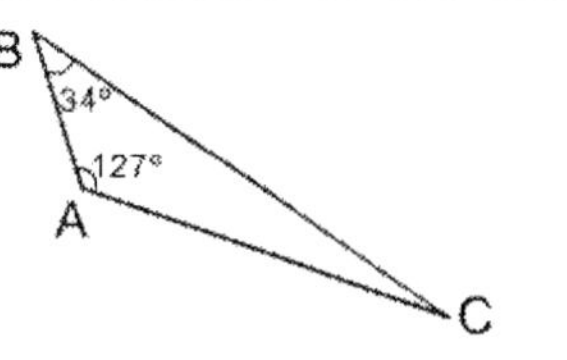

A. 25 °   B. 17°   C. 19°

12. An angle of 90° is called?

A. obtuse   B. right   C. acute

13. Classify the following triangle.

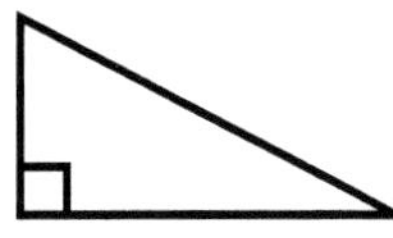

A. right   B. obtuse   C. equilateral

14. Which shape has 2 pairs of parallel sides, and 2 pairs of angles that are equal?

A.   B.    C. 

15. Which pair of shapes look congruent?

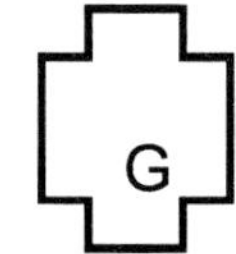
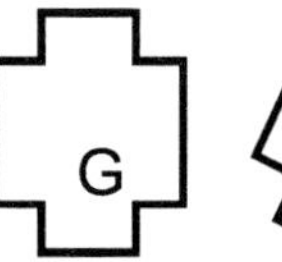

A. F and H   B. G and H   C. E and F

16. Look at the shapes. Choose flip, slide or turn.

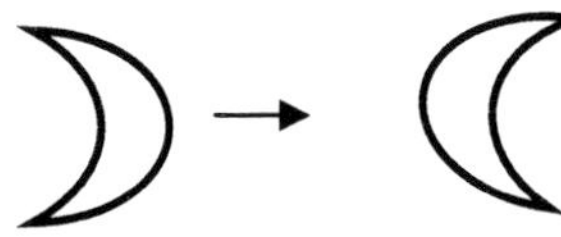

A. flip   B. slide   C. turn

17. How many of the following pairs of line are intersecting?

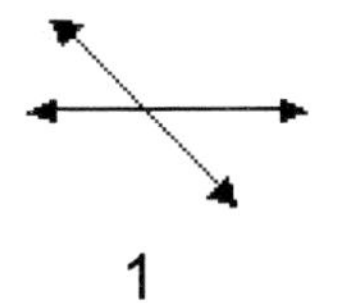
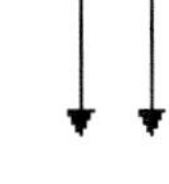

1   2   3

A. 1   B. 2   C. 3

18. How many edges?

A. 8   B. 11   C. 12

19. What 3D figure could be made from this net?

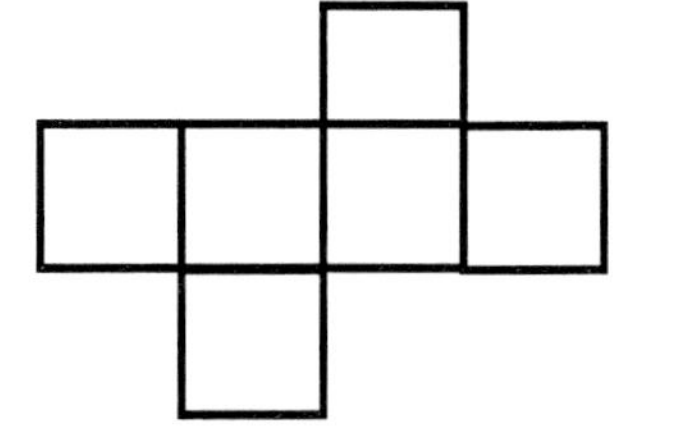

A. cylinder   B. cube   C. pyramid

20. What 3D figure does this object look like?

A. cube   B. sphere   C. cylinder

Name ______________

Math Test: Ordered Pairs

A. Place the ordered pairs on the grid.

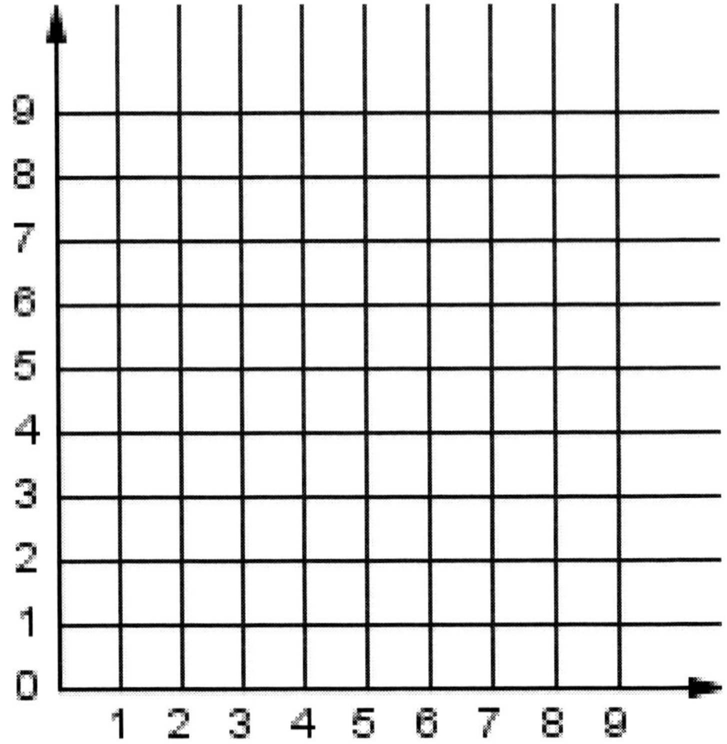

A. (5, 6)	B. (2, 2)
C. (1, 7)	D. (5, 9)
E. (9, 2)	F. (8, 1)
G. (6, 9)	H. (3, 8)
I. (0, 4)	J. (4, 3)

Number Correct: /10

Name ______________

Math Test: Ordered Pairs

A. Place the ordered pairs on the grid.

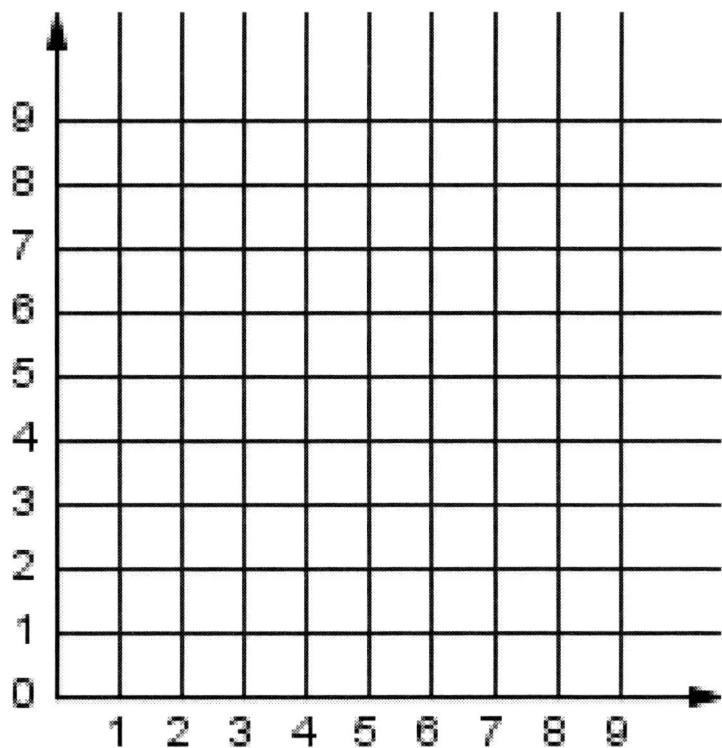

A. (9, 9)	B. (1,1)
C. (4, 8)	D. (3, 6)
E. (1, 2)	F. (6, 8)
G. (6, 5)	H. (7, 8)
I. (8, 1)	J. (3, 3)

Number Correct: /10

Measurement

Tests

Name ________________

Math Test: Telling Time

A. Write the time.

1.  ____ : ____	2.  ____ : ____	3.  ____ : ____	4.  ____ : ____
5.  ____ : ____	6.  ____ : ____	7.  ____ : ____	8.  ____ : ____
9.  ____ : ____	10.  ____ : ____	11.  ____ : ____	12.  ____ : ____

Name ______________

Math Test: Telling Time

A. Write the time.

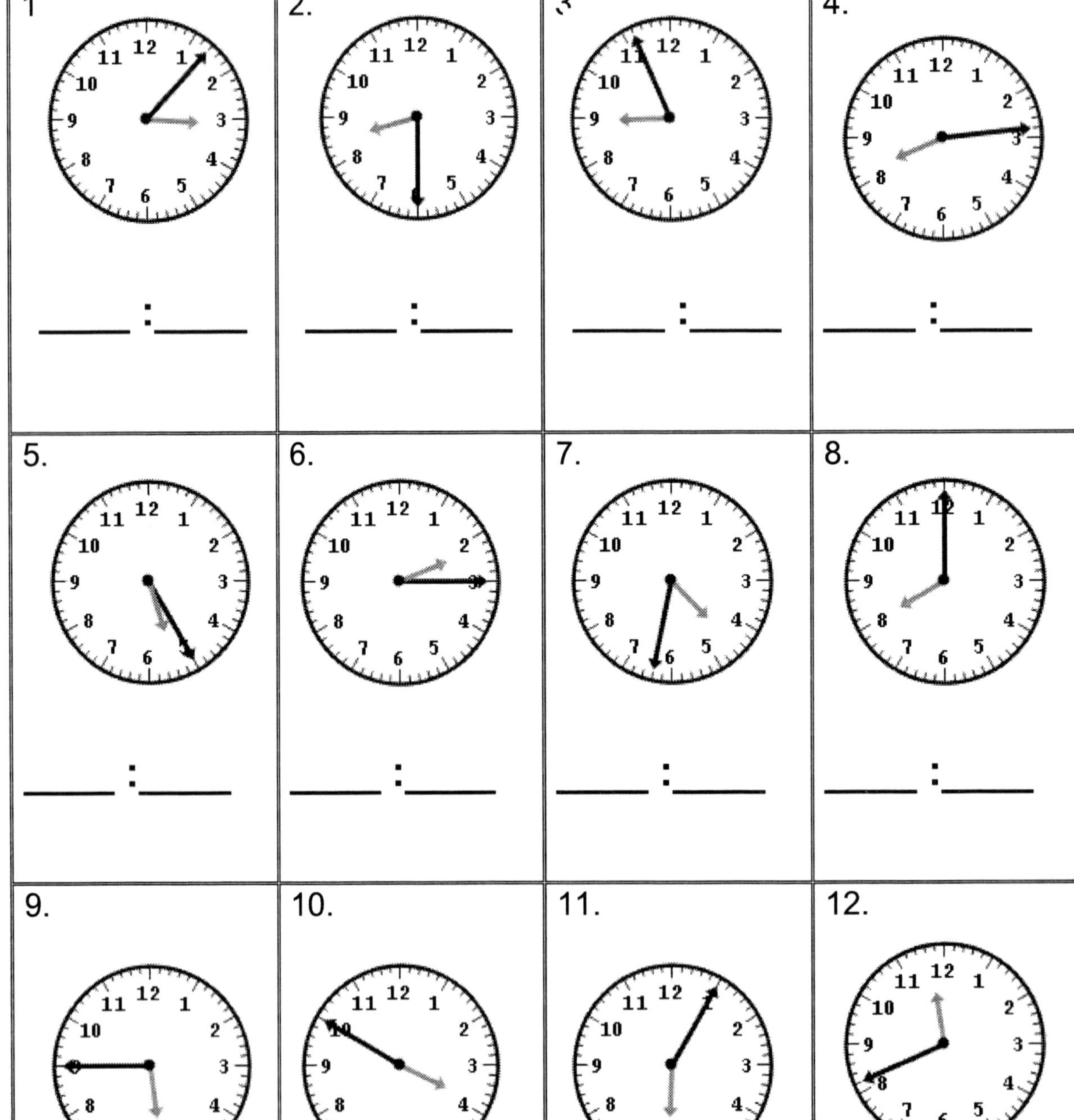

Name ______________

Math Test: Calculate the Elapsed Time

1. The time is 4:14 a.m. What time will it be in 40 minutes?
2. The time is 2:15 p.m. What time will it be in 1 hour and 10 minutes?
3. The time is 9:34 p.m. What time will it be in 2 hours?
4. The time is 11:43 p.m. What time will it be in 22 minutes?
5. The time is 7:05 p.m. What time will it be in 1 hour and 30 minutes?
6. The time is 9:34 p.m. What time will it be in 2 hours and 10 minutes?
7. The time is 9:34 a.m. What time will it be in 75 minutes?
8. The time is 10:07 p.m. What time will it be in 120 minutes?
9. The time is12:34 a.m. What time will it be in 85 minutes?
10. The time is 6:05 p.m. What time will it be in 3 hours and 25 minutes?

Name ________________

Math Test: Time

Circle the letter next to the correct answer.

1. How many minutes in 6 hours?

A. 310 minutes
B. 360 minutes
C. 390 minutes

2. How many days in 8 weeks?

A. 28 days
B. 80 days
C. 56 days

3. What is another way to write this date: July 13, 2006?

A. 2006, 02, 07
B. 2006 09, 02
C. 2006, 07, 13

4. Each week Ben plays hockey for 320 minutes. How many hours and minutes does he play hockey?

A. 2 hours 40 minutes
B. 5 hours 20 minutes
C. 3 hours 30 minutes

5. About how long would it take to brush your teeth?

A. 4 minutes
B. 4 hours
C. 4 days

6. How many decades in 5 centuries?

A. 50
B. 60
C. 70

7. How many months in 3 years?

A. 24 months
B. 36 months
C. 48 months

8. How many weeks in a year?

A. 52 weeks
B. 35 weeks
C. 100 weeks

9. What is the year 1 decade after 1968?

A. 1975
B. 1978
C. 1969

10. What time will it be 120 minutes later?

A. 9:30
B. 8:30
C. 5:30

Relating Units of Measurement #1

Name ______________

1) 30 cm = ___ mm	2) 30 dm = ___ cm	3) 4 km = ____ m
4) 5 m = ____ cm	5) 600 cm = ___ m	6) 22 cm = ___ mm
7) 70 mm = ___ cm	8) 40 dm = ___ cm	9) 14 m = ____ cm
10) 8000 m =___km	11) 500 cm = ___ m	12) 80 dm = ___ cm
13) 200 cm = ___ m	14) 6 km = ___ m	15) 30 m = ____ cm
16) 10 dm = ___ cm	17) 9 m = ____ cm	18) 100 cm = ___ m

Relating Units of Measurement #2

Name ____________

1) 70 dm = ____ cm	2) 10 cm = ____ mm	3) 5 km = ____ m
4) 900 cm = ____ m	5) 20 m = ____ cm	6) 9 cm = ____ mm
7) 20 dm = ____ cm	8) 30 mm = ____ cm	9) 76 m = ____ cm
10) 700 cm = ____ m	11) 45 cm = ____ mm	12) 90 km = ____ m
13) 50 dm = ____ cm	14) 800 cm = ____ m	15) 2 m = ____ cm
16) 4000 m =____ km	17) 10 dm = ____ cm	18) 400 cm = ____ m

Name _______________

Math Test: Units of Measure

A. State the best unit measure to measure:

1. the length of a swimming pool
2. the width of a finger
3. the distance between two towns
4. the height of a building
5. a drop of rain
6. the length of a football field
7. the height of a large carton of milk
8. the weight of 5 books
9. the time it takes to drink water from a fountain
10. the length of an ant
11. a cup of coffee
12. the weight of a car
13. the capacity of a bath tub
14. the time you spend in school
15. the temperature outside

Number Correct: /15

Name ______________

Math Test: Perimeter

A. Find the perimeters of each figure.

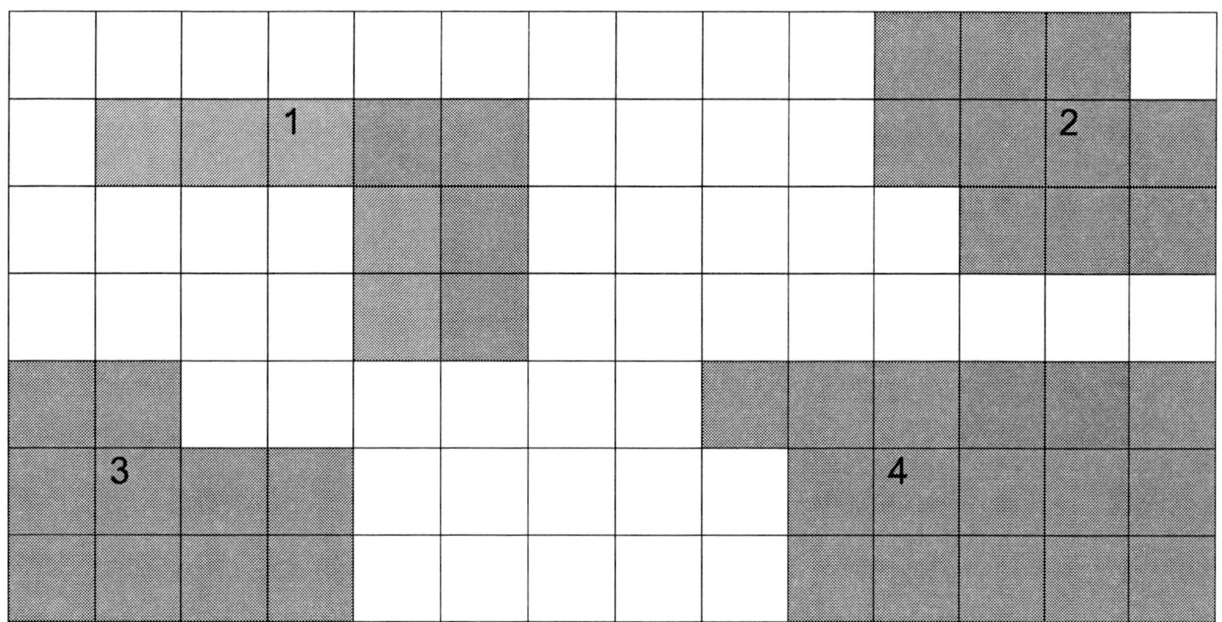

1. The perimeter of figure 1 is: __________units.

2. The perimeter of figure 2 is: __________units.

3. The perimeter of figure 3 is: __________units.

4. The perimeter of figure 4 is: __________units.

B. Solve:

Both of these figures have 10 squares.
Circle the figure with longest perimeter.

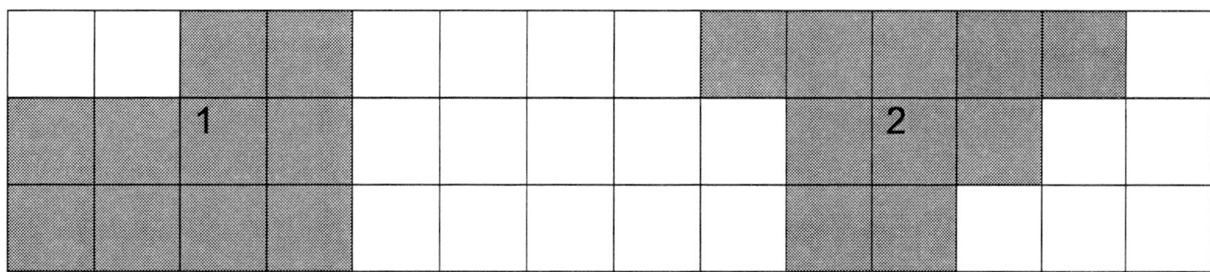

Name ______________

Math Test: Area

A. Find the areas of each figure.

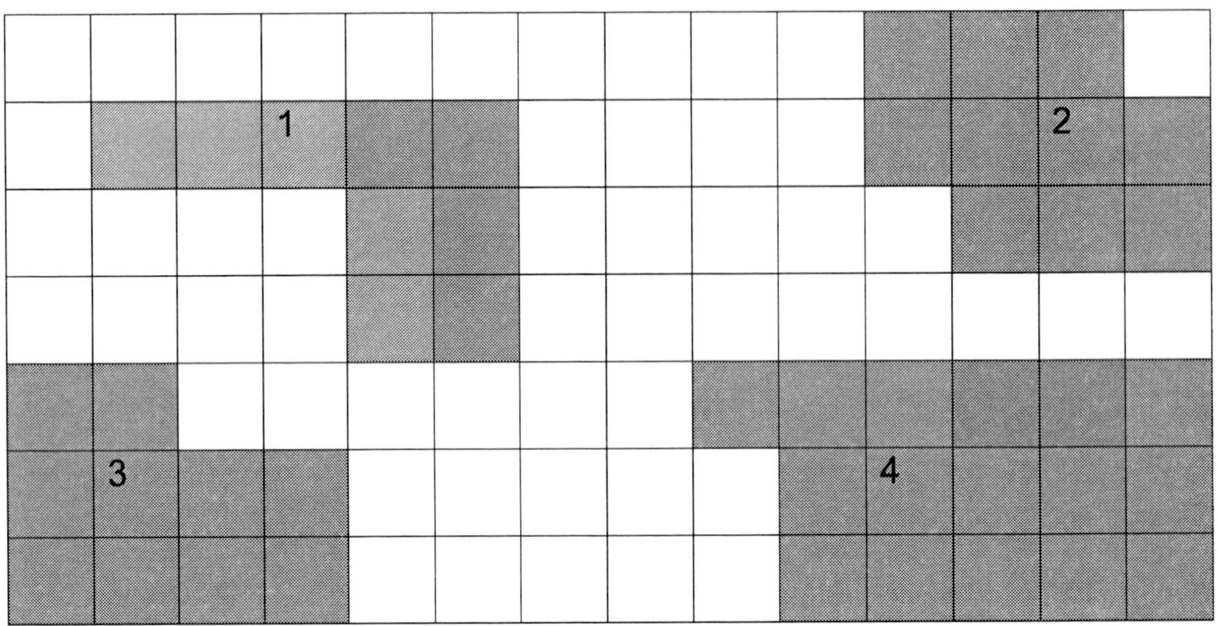

1. The area of figure 1 is: __________units^2.

2. The area of figure 2 is: __________units^2.

3. The area of figure 3 is: __________units^2.

4. The area of figure 4 is: __________units^2.

B. Solve:

1. Both of these figures have 10 squares.
   Circle the figure with smallest area.

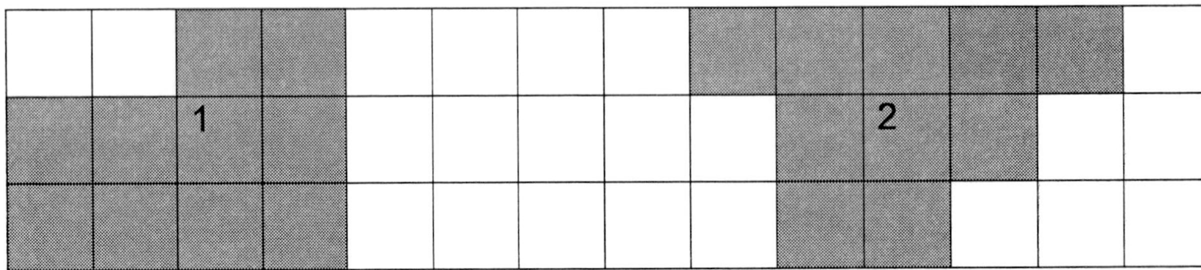

Name ______________

Math Test: Perimeter and Area

A. Find the perimeter and area for each figure.

1.

Perimeter= _______ units

Area= ______square units

2.

Perimeter= _______ units

Area= ______square units

3.

Perimeter= _______ units

Area= ______square units

4.

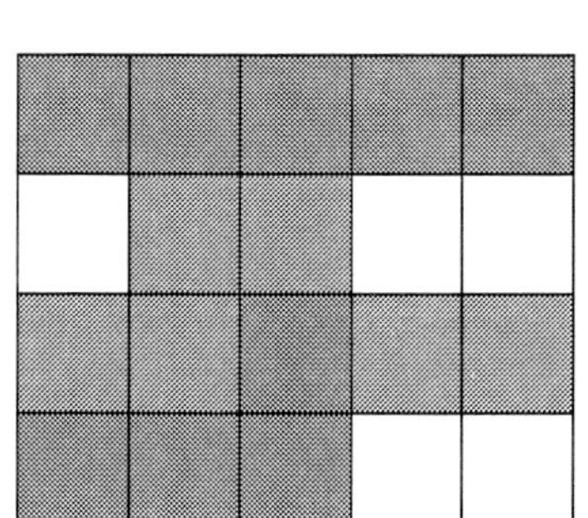

Perimeter= _______ units

Area= ______square units

5.

Perimeter= _______ units

Area= ______square units

6.

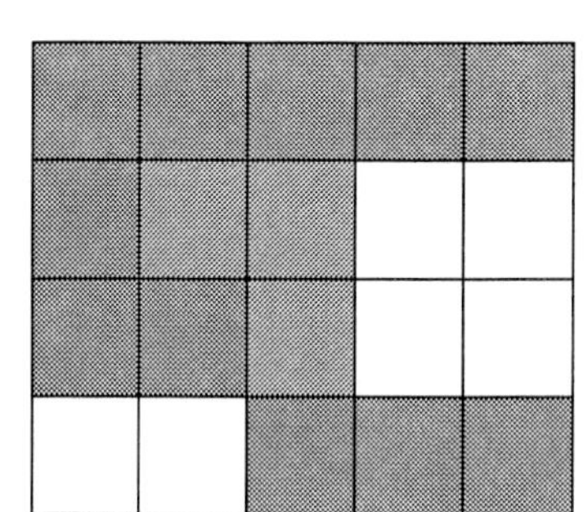

Perimeter= _______ units

Area= ______square units

7.

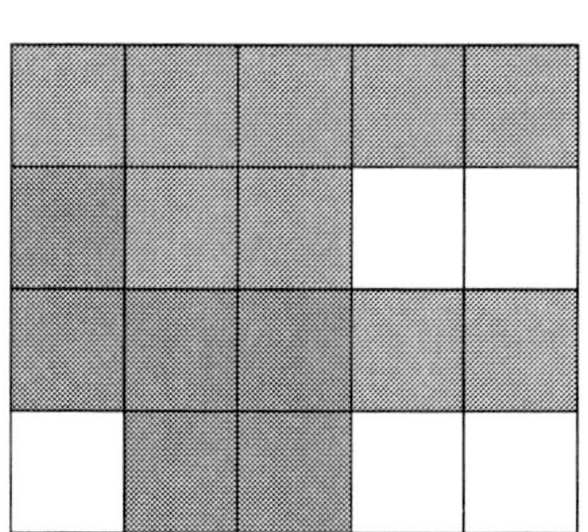

Perimeter= _______ units

Area= ______square units

8.

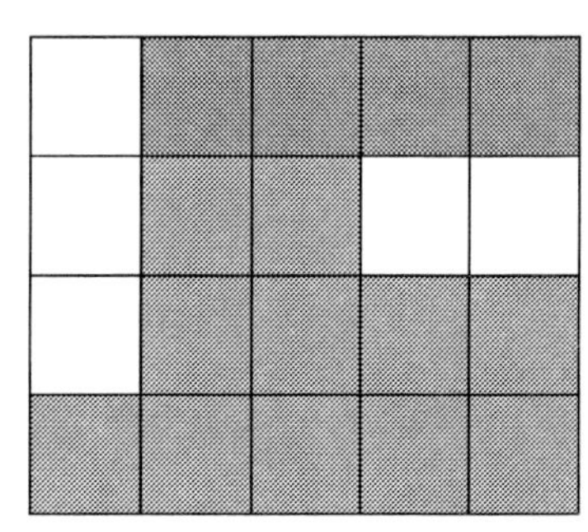

Perimeter= _______ units

Area= ______square units

9.

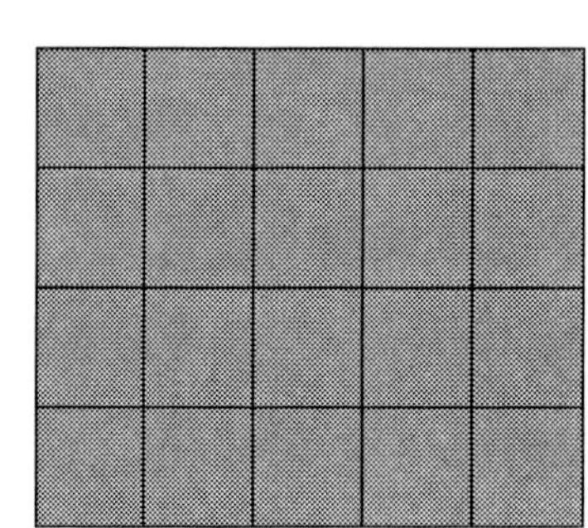

Perimeter= _______ units

Area= ______square units

Name ______________

Math Test: Perimeter

A. Calculate the perimeters.

1. Find the perimeter of the rectangle.

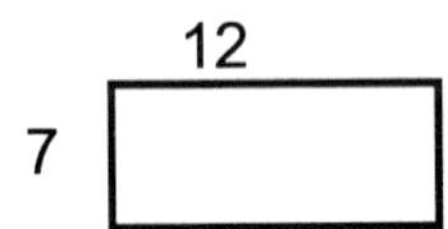

2. Find the perimeter of the square.

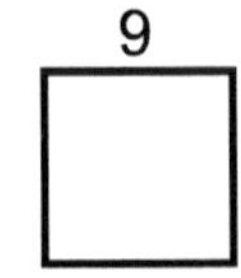

3. Find the perimeter of the square.

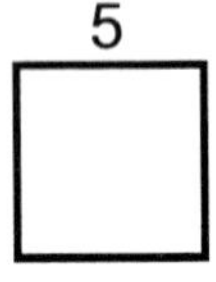

4. Find the perimeter of the trapezoid.

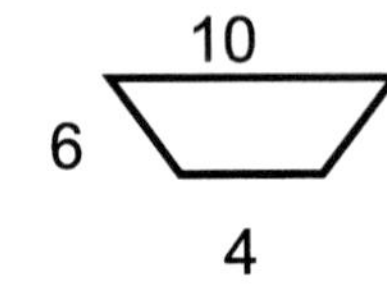

5. Find the perimeter of the rectangle.

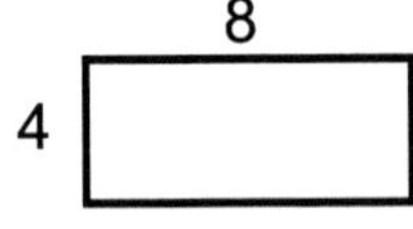

6. Find the perimeter of the equilateral triangle.

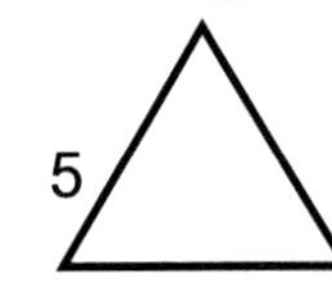

7. Find the perimeter of the octagon.

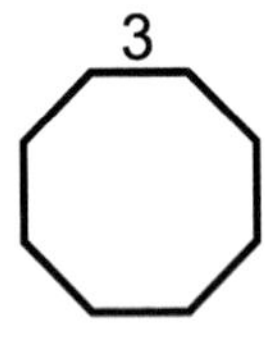

8. Find the perimeter of the trapezoid.

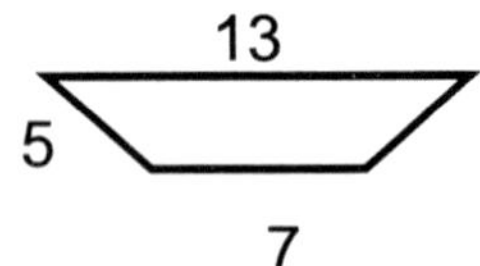

9. Find the perimeter of the equilateral triangle.

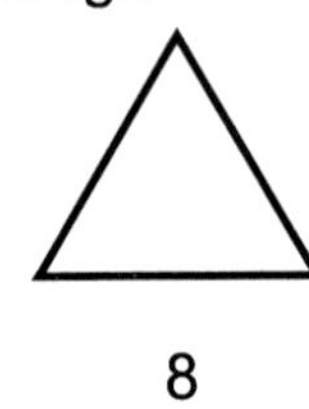

10. Find the perimeter of the square.

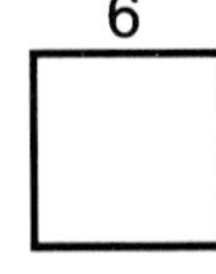

Number Correct: /10

Name ______________

Math Test: Perimeter

A. Find the perimeter each irregular polygon.

1.

7 m
4 m
4 m
2 m
3 m
2 m
2 m
2 m

2.

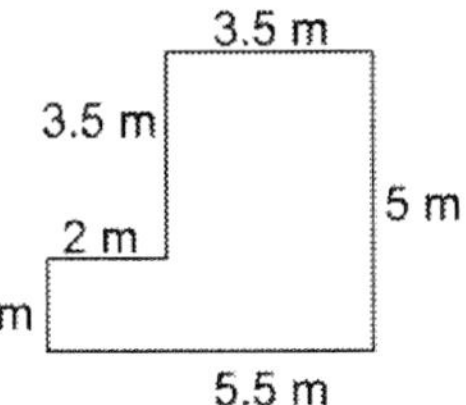

3.

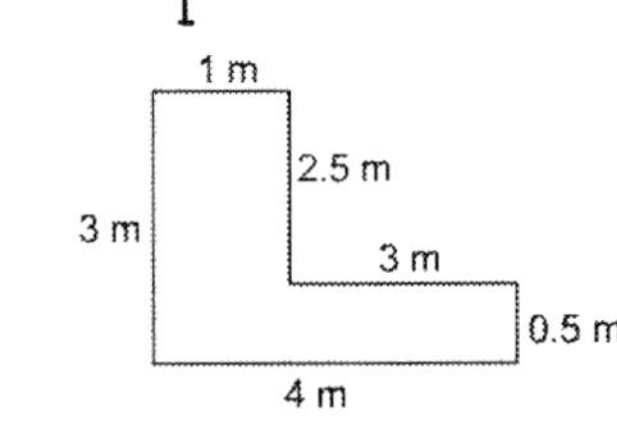

4.

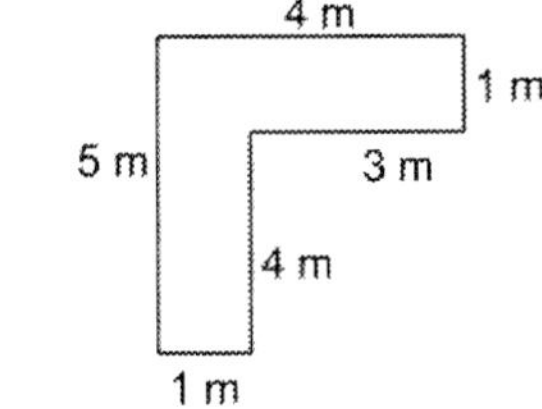

5.

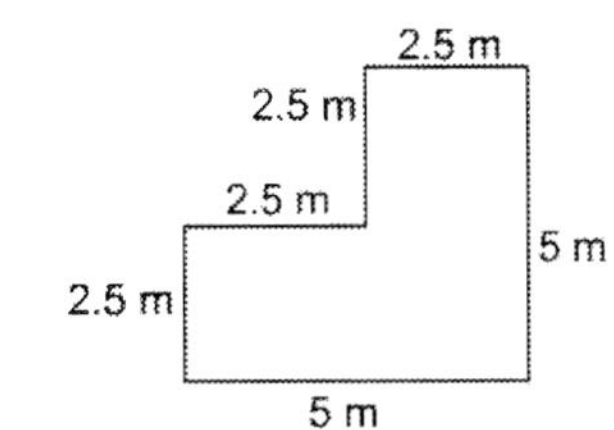

Number Correct: /5

Name ___________

Math Test: Perimeter and Area

A. Find the perimeter and area of the following figures.

1. Show your work:

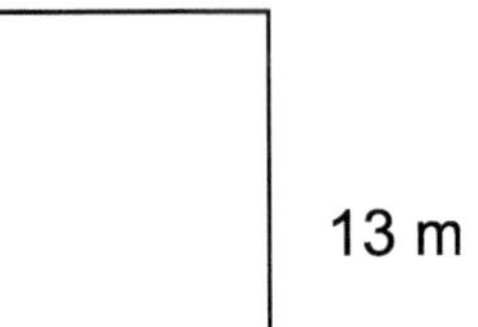

13 m

8 m

Perimeter = ________

Area = ________

2. Show your work:

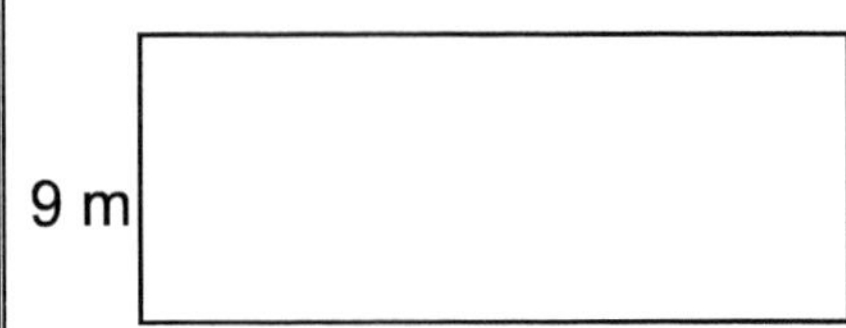

9 m

17 m

Perimeter = ________

Area = ________

3. Show your work:

18 m

7 m

Perimeter = ________

Area = ________

4. Show your work:

16 m

6 m

Perimeter = ________

Area = ________

5. Show your work:

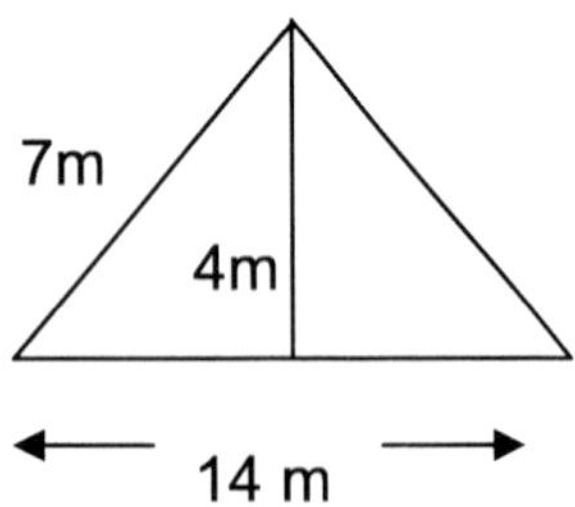

14 m

Perimeter = ________

Area = ________

6. Show your work:

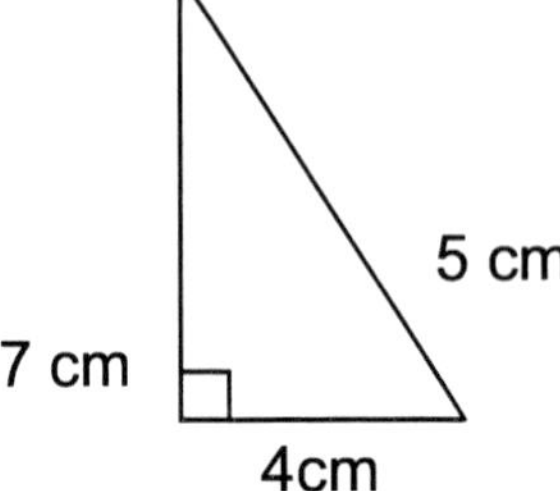

Perimeter = ________

Area = ________

Name ________________

Math Test: Area

A. Calculate the area for each parallelogram. Remember to write the proper units.

Area = b x h

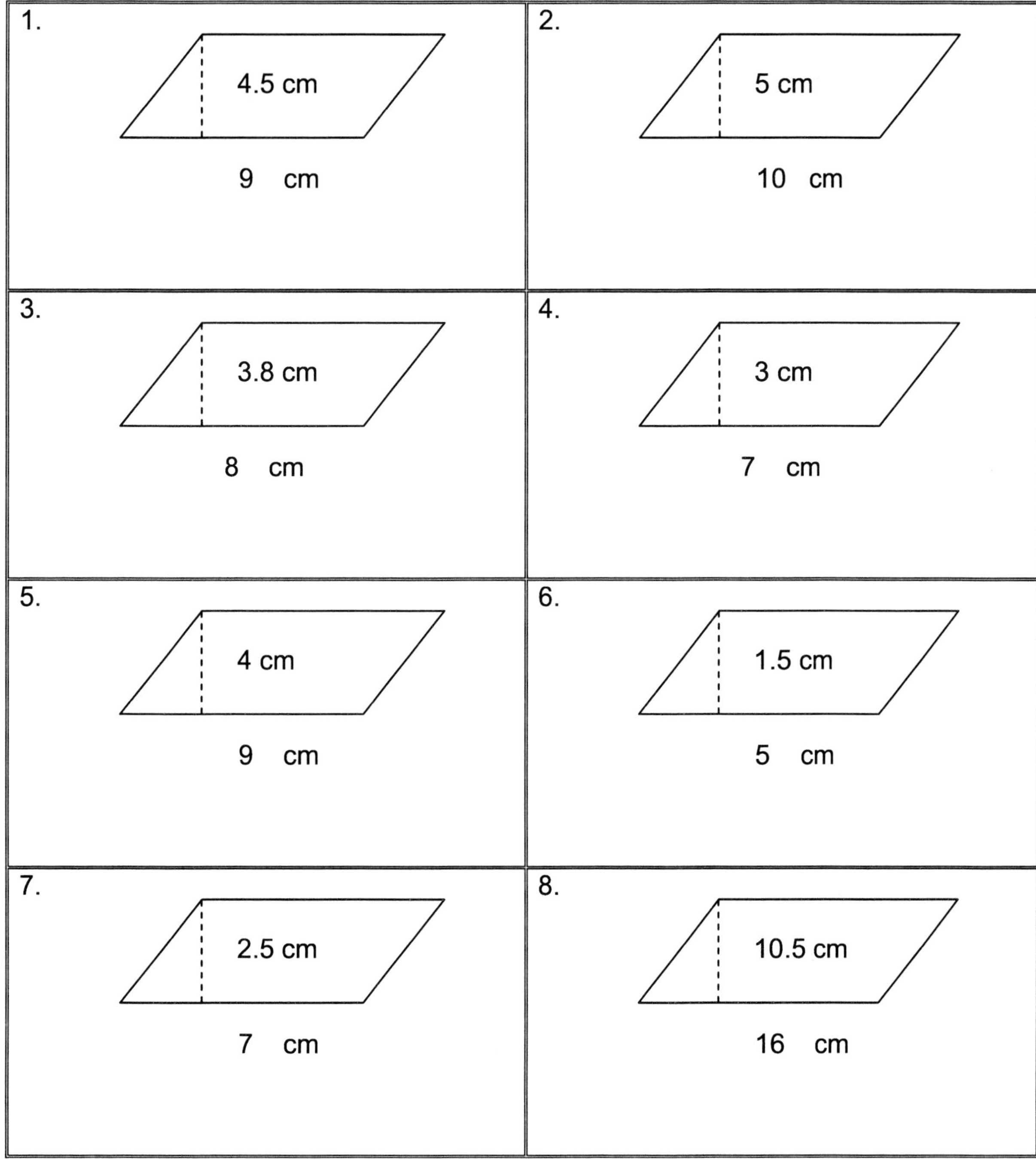

Name ________________

Math Journal

How is measurement used in everyday life?

Name ______________

Math Test: Volume

A. Calculate the volume of each prism.

1.

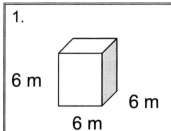

6 m

6 m

6 m

Volume =

2.

6 m

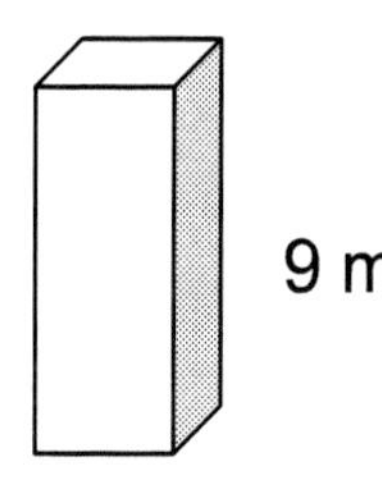

9 m

6 m

Volume=

3.

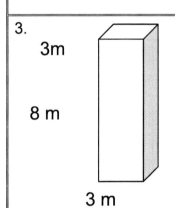

3m

8 m

3 m

Volume =

4. 3 mm

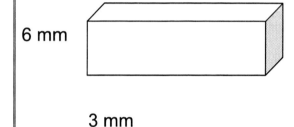

6 mm

3 mm

Volume =

5.

9 m

4 m

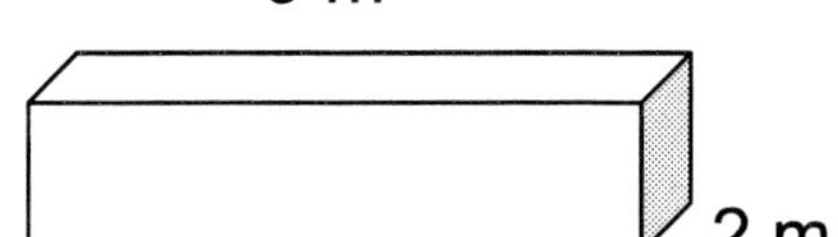

2 m

Volume =

6. 7 m

7 m

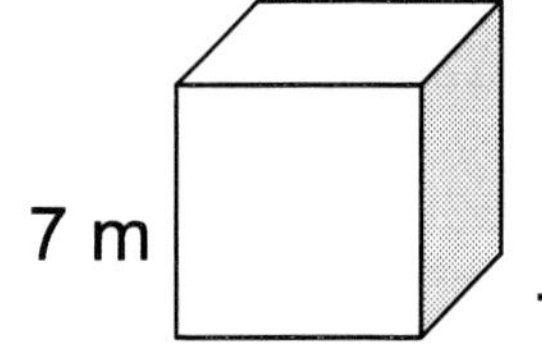

7 m

Volume =

Name ______________

Math Test: Measurement Grade 4

Circle the letter next to the correct answer.

1. What temperature is it most likely to be if you are swimming in a lake during the summer? A. 10 ° B. 30° C. 0°	2. Which is the shortest? A. 10 cm B. 10 dm C. 10 m
3. What is the best estimate for the length of a worm? A. 6 cm B. 6 m C. 6 km	4. What is the perimeter of the rectangle? 7 14 A. 38 B. 21 C. 42
5. Which is the longest? A. 10 cm B. 10 dm C. 10 m	6. What unit of mass would you use to measure the mass of a bike? A. mg B. kg C. g
7. What is the area of the rectangle? 6 13 A. 78 B. 38 C. 19	8. What unit of measure would you use to measure the capacity of a juice carton? A. litres B. grams C. metres
9. What is the best estimate for the mass of two books? A. 1 kg B. 1 g C. 1000 kg	10. What is the best estimate for the capacity of a cup of hot chocolate? A. 200 l B. 200 ml C. 2 ml

Name ______________

Math Test: Measurement Grade 4

Circle the letter next to the correct answer.

11. What temperature is it most likely to be if you are outside playing in the snow? A. – 5 ° B. 12° C. 5°	12. Which is the shortest? A. 100 cm B. 100 dm C. 100 m
13. What is the best estimate for the length of a teacher's desk? A. 1 cm B. 1 m C. 1 km	14. What is the perimeter of the hexagon? 5 A. 20 B. 25 C. 30
15. Choose the shape that would be the best unit to measure this area. A. ◯ B. △ C. □	16. What is the area in square metres of a pool 12 m X 20 m? A. 120 m^2 B. 240 m^2 C. 20 m
17. A rectangle is 4 cm long and 7cm wide. What is the perimeter if the length decreased by 1 cm? A. 28 cm B. 21 cm C. 20 cm	18. Each side of a square measures 6 cm. What would the perimeter be if the sides were to increase by 2 cm each? A. 24 cm B. 32 cm C. 28 cm
19. A rectangle has an area of 56 cm^2. The width is 8 cm. What is the length? A. 8 cm B. 7cm C. 15 cm	20. How many decimeters in 200 cm? A. 2 dm B. 20 dm C. 2000 dm

Name ______________

Math Test: Measurement Grade 5

Circle the letter next to the correct answer.

1. David lives 750 000 cm from his school. How many metres is that? A. 75 m B. 7.5 m C. 750 m	2. What object is shorter in length than a decimetre? A. an ant B. a bus C. a metre stick
3. What is the best estimate for the length of a student's desk? A. 86 cm B. 8 m C. 8 km	4. What is the perimeter of the figure? 4 m, 1 m, 3 m, 5 m, 4 m, 1 m A. 18 m B. 20 m C. 22 m
5. Which is the longest? A. 5 cm B. 5 dm C. 5 m	6. What unit of mass would you use to measure the mass of a car? A. tonnes B. kilograms C. grams
7. What is the area of the rectangle? 8, 12 A. 78 B. 38 C. 96	8. What unit of measure would you use to measure the capacity of a milk carton? A. litres B. grams C. metres
9. How many metres in 45 km? A. 45 000 m B. 450 m C. 4.5 m	10. How many kilograms in 600 grams? A. 60 kg B. 6 kg C. 0.06 kg

Name ______________

Math Test: Measurement Grade 5

Circle the letter next to the correct answer.

11. What temperature would it most likely be if you are skating on a pond? A. – 5 ° B. 12° C. 5°	12. Jane's stride is about 60 cm. It took her 9.5 strides to measure the side of her house. About how long is the side of her house? A. 600 cm B. 570 dm C. 570 cm
13. A round table has a width of 1.5 m. Which is the best estimate of the table's circumference? A. 4.7 m B. 8 m C. 3.5 m	14. What is the perimeter of the hexagon? 15 A. 80 B. 90 C. 70
15. Choose the shape that would be the best unit to measure this area. A. B. C.	16. What is the area in square metres of a pool 12 m X 20 m? A. 240 m^2 B. 64 m^2 C. 120 m
17. A rectangle is 12 cm long and 8 cm wide. What is the perimeter if the length decreased by 2 cm? A. 112 cm B. 36 cm cm C. 120 cm	18. Each side of a square measures 9 cm. What would the perimeter be if the sides were to increase by 3 cm? A. 24 cm B. 32 cm C. 48 cm
19. A rectangle has an area of 176 cm^2. The width is 11 cm. What is the length? A. 18 cm B. 17cm C. 16 cm	2 0 . How many mm in 500 cm? A. 50 mm B. 5000 mm C. 50 000 mm

Name _______________

Math Test: Measurement Grade 6

Circle the letter next to the correct answer.

1. Lucy lives 84 000 cm from her school. How many metres is that? A. 84 m B. 8.4 m C. 840 m	2. A box of cookies weighs 520 grams. If Jane bought two boxes for the class party, what is the combined weight of both boxes of cookies in kilograms? A. 0.1040 kg B. 1.040 kg C. 104.0 kg
3. What is the best estimate of the length of a bicycle? A. 1 m B. 3 m C. 8 m	1. How litres in 5321 ml? A. 5.321 l B. 53.21 l C. 0.5321 l
2. If the perimeter of a hexagon is 19.2 m, what is the length of each side? A. 3 m B. 4.3 m C. 3.2 m	3. What unit of length would you use to measure the thickness of a loonie? A. mm B. cm C. dm
4. What is the area of the rectangle to the nearest ones? 8.9 m 12 m A. 109 m^2 B. 107 m^2 C. 106.8 m^2	5. What is the best estimate of the distance between the cities of Toronto and London? A. 2 dm B. 200 km C. 1000 m
6. What is the best estimate for the mass of two books? A. 1 kg B. 1 g C. 1000 kg	7. What is the area of a square with sides 9 m? A. 36 m^2 B. 81 m^2 C. 18 m^2

Name ______________

Math Test: Measurement Grade 6

Circle the letter next to the correct answer.

11. If a circle has a radius of 6 cm, what is the circumference to the nearest tenth? A. 18.84 cm B. 18 cm C. 18.8 cm	12. What is the best unit of measurement to weigh a pack of gum? A. g B. kg C. mg
13. A round table has a width of 3 m. Which is the best estimate of the table's circumference to the nearest tenth? A. 9.4 m B. 8 m C. 3.5 m	14. What is the perimeter of the hexagon? 25.6 m A. 153.6 m B. 150.5 m C. 148 m
15. Choose the shape that would be the best unit to measure this area. A. (circle) B. (square) C. (triangle)	16. What is the area in square metres of a pool 27.2 m X 9 m? A. 240 m^2 B. 244.8 m^2 C. 234.8 m
17. A rectangle is 18 cm long and 9.2 cm wide. What is the perimeter if the length decreased by 2 cm? A. 40 cm B. 50.4 cm C. 42. 4 cm	18. Each side of a hexagon measures 6 cm. What would the perimeter be if the sides were to increase by 3 cm each? A. 72 cm B. 54 cm C. 32 cm
19. A rectangle has an area of 216 cm^2 . The width is 12 cm. What is the length? A. 18 cm B. 17cm C. 16 cm	20. A circle has a diameter of 16 m. What is the radius? A. 20 m B. 8 m C. 10 m

Data Management Tests

Name ________________

Data Management #1

A. Complete the chart.

Set of Data	Mean	Range	Median	Mode
1. 28, 21, 3, 10, 6, 9, and 21				
2. 24, 3, 17, 23, 19, 19, and 7				
3. 13, 4,9 23, 24, and 23				
4. 12, 21, 12, 7,and 8				
5. 7, 19, 28, 8, and 28				

Data Management #2

A. Complete the chart.

Set of Data	Mean	Range	Median	Mode
1. 5, 25, 14, 25, 25, 2, and 2				
2. 7, 6, 2, 6, 28, 4, and 10				
3. 15, 27, 28, 5, and 5				
4. 22, 25, 17, 22, and 4				
5. 7, 28, 7, 6, and 7				

Name ______________

Data Management #3

A. Complete the chart.

Set of Data	Mean	Range	Median	Mode
1. 16, 17, 15, 16, and 11				
2. 27, 15, 15, 2, and 26				
3. 3, 3, 6, 29, 21, 20, and 2				
4. 14, 11, 14, 19, and 22				
5. 5, 8, 28, 5, and 19				

Data Management #4

Name ______________

A. Complete the chart.

Set of Data	Mean	Range	Median	Mode
1. 18, 12, 11, 2, and 17				
2. 15, 25, 25, 10, and 25				
3. 6, 28, 10, 10, and 11				
4. 24, 24, 2, 22, and 23				
5. 7, 17, 26, 17, and 3				

Name _______________

Math Test: Data Management

Pop Tabs Collected

	Week 1	Week 2	Week 3
Monday	23	20	13
Tuesday	11	24	6
Wednesday	30	8	13
Thursday	18	17	28
Friday	11	15	8

	Show your work.
1. How many pop tabs were collected over the past three Tuesdays?	
2. During Week 3, how many fewer pop tabs were collected on Monday than Friday?	
3. What was the range of the number of pop tabs collected in Week 2?	
4. How many pop tabs were collected on Wednesday of Week 2?	
5. How many pop tabs were collected over the past three Thursdays?	
6. What was the median of the number of pop tabs collected in Week 2?	
7. What was the total number of pop tabs collected in week 3?	
8. How many more pop tabs were collected on the Thursdays than on the Tuesdays?	
9. What was the mode of the number of pop tabs collected in week 1?	
10. What was the mean of the number of pop tabs collected in Week 1?	

Name ______________

Math Test: Data Management

Here is a double bar graph of how much money two classes collected for a class trip.

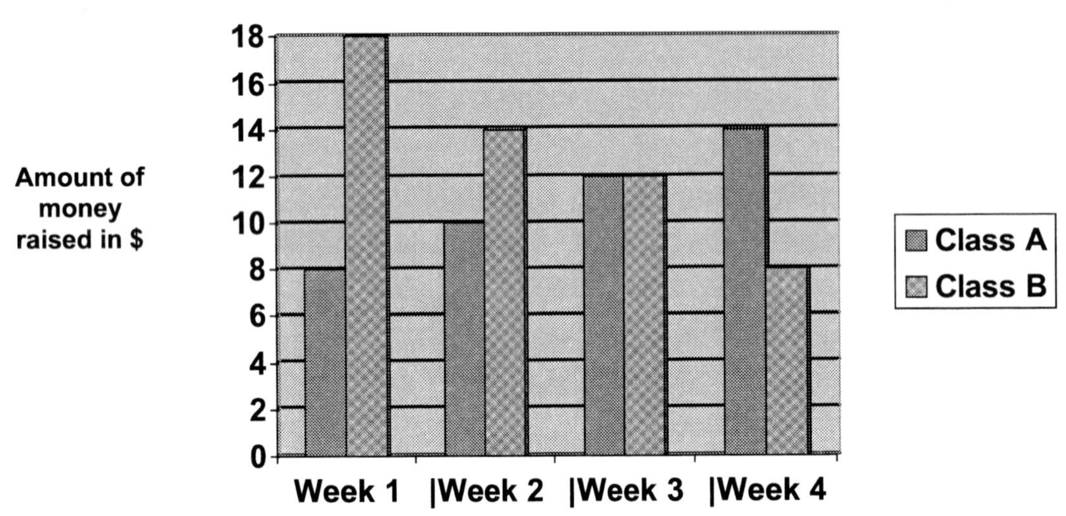

Write 10 things you know from the graph.

1. ______________________________

2. ______________________________

3. ______________________________

4. ______________________________

5. ______________________________

6. ______________________________

7. ______________________________

8. ______________________________

9. ______________________________

10. ______________________________

Name ______________

Math Test: Constructing Bar Graphs

A. Use the data from the tally chart to make a bar graph. Make sure you label!

Subject	Tally	Number
Reading	~~IIII~~ IIII	
Math	~~IIII~~ II	
Art	~~IIII~~ II	
Science	IIII	

____

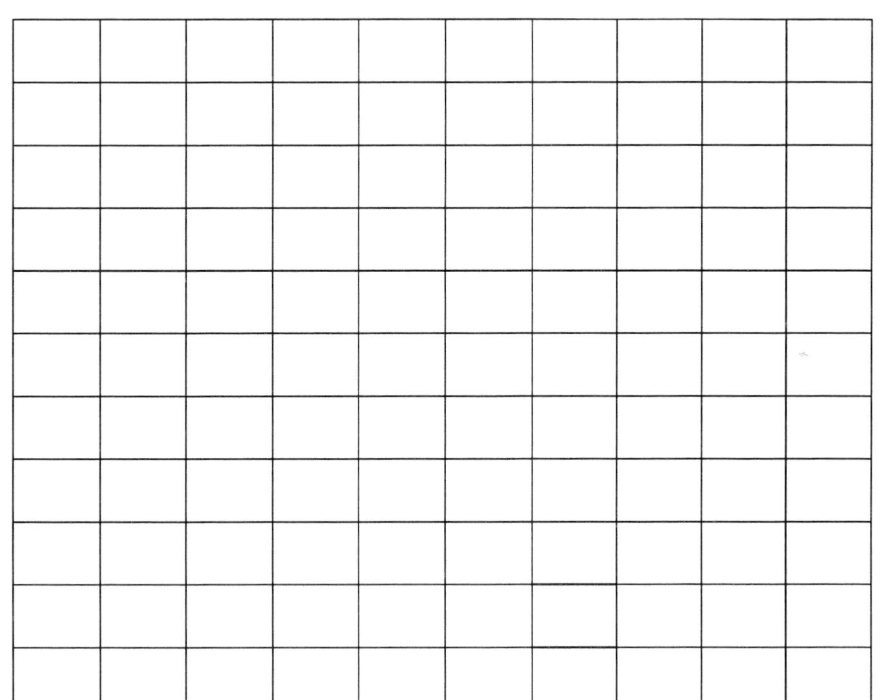

____

Write three things you know from the graph.

1. ______________________

2. ______________________

3. ______________________

Name ____________

Math Test: Data Management

Here are the results of a Favourite Sports Survey. Football got 25 votes, baseball 75, basketball 20, hockey 70 and tennis 10 votes.

1. Create a tally chart to show the information.

Football	Baseball	Basketball	Hockey	Tennis

2. Complete the horizontal bar graph.

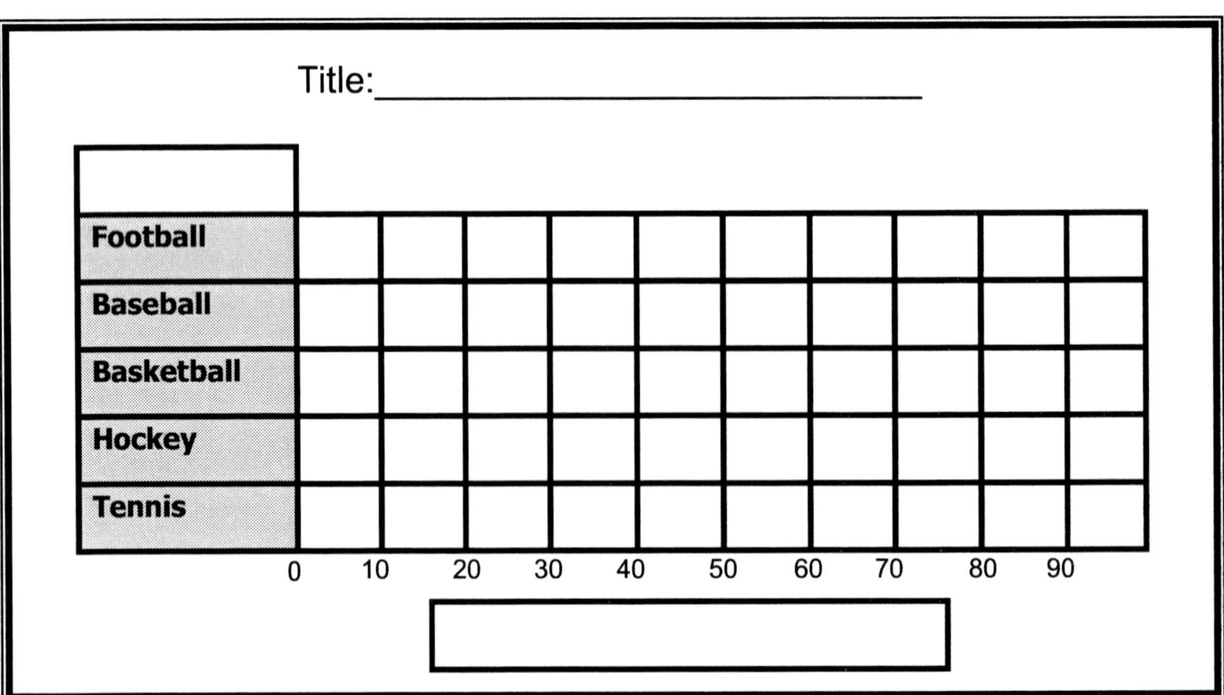

3. What was the most popular sport? ________________

4. What was the least popular sport? ________________

5. How many people voted in the survey? ________________

Name ______________

Math Test: Data Management

Read the pictograph to answer the questions about the results of the survey.

Favourite Ice Cream Flavour

Rocky Road	
Chocolate Chip	
Vanilla	
Chocolate	

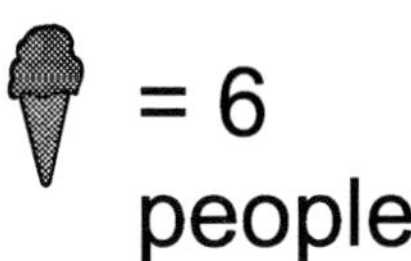

1. What flavour is the most popular? ____________________

2. What flavour is the least popular? ____________________

3. How many students chose vanilla? ____________________

4. How many fewer students chose chocolate over vanilla? ________________

5. How many students chose either rocky road or vanilla? ____________________

6. How many people chose chocolate? ____________________

7. How many people chose chocolate chip? _________________

8. What is the ratio of rocky road to chocolate chip? __________________________

9. What is the ratio of vanilla to chocolate? ____________________________

10. How many people voted altogether? _________________________

Name ______________

Math Test: Data Management

A. Complete the line graph using the data from the table.

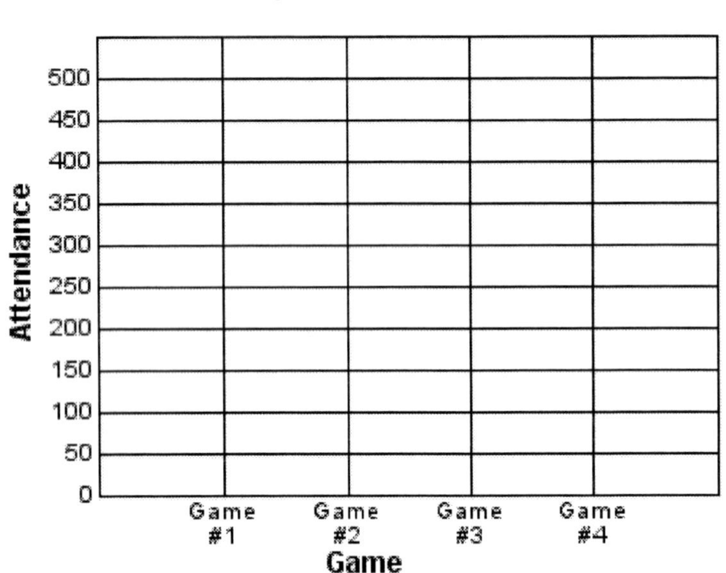

Game	# of People
Game # 1	100
Game # 2	250
Game # 3	300
Game #4	450

B. Write five things you know from the graph.

1. ______________________________

2. ______________________________

3. ______________________________

4. ______________________________

5. ______________________________

C. Explain why a line graph is best to show this data.

____

____

Name ______________

Math Test: Data Management

A. Complete a line graph using the data from the table.

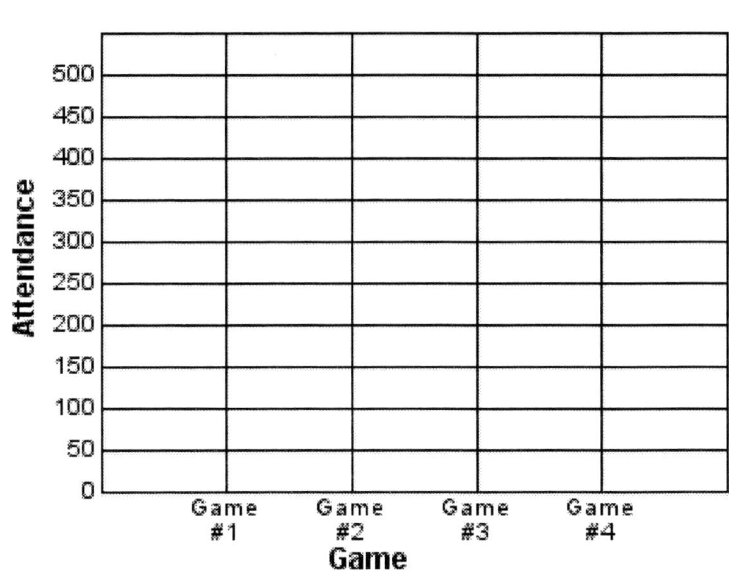

Game	# of People
Game # 1	500
Game # 2	350
Game # 3	400
Game #4	150

B. Answer the following questions.

1. Which game had the most attendance? ____________________
2. Which game had the least attendance? ____________________
3. Which game(s) did not have an attendance of at least 350 people? _______________
4. Which game(s) had an attendance of less than 400 people? __________________
5. How many people altogether attended all four games? ____________________
6. Which game showed a decrease of 150 people from the week before? ___________
7. Which game showed the biggest decrease from the week before? _____________

Name ______________

Math Journal

How is math used is everyday life?

Name ______________

Math Journal

What do you like about math?

What don't you like about math?

Name ______________

Math Journal

The one thing I need to work on in math is.....

Name ______________

Math Journal

In math I am good at ....

Name ______________

Math Journal

What strategies do you use to figure out math problems?

Name ________________

Math Rubric

	Did I solve it?	Did I show my math work?
Full Speed Ahead!	➢ I completed the problem independently with no mistakes.	➢ I showed all of the steps to solve the problem. ➢ I wrote math language to explain my thinking.
Keep Going!	➢ I completed the problem independently, but made a few mistakes.	➢ I showed most of the steps to solve the problem. ➢ I used some math language to explain my thinking.
Slow Down!	➢ I tried to do the problem, but I need help.	➢ I showed a few steps to solve the problem. ➢ I used a little math language to explain my thinking.
Stop!	➢ I didn't try the problem.	➢ I did not show my math work.

Name ______________

Anecdotal Observations

Math Focus ____

	Strengths	Weaknesses
Completion of Daily Math Work		
Understanding of Math Concepts		
Application of Skills Taught		
Application of Math Terminology to Explain Ideas		
Attendance to Task		

MATH TESTS CLASS RECORD

Student Name															

Math Test: Addition and Subtraction

A. Add or subtract the following.

1. 909 - 176 733	2. 242 - 140 102	3. 329 + 583 912	4. 833 - 475 358	5. 315 + 338 653
6. 127 + 405 532	7. 435 - 127 308	8. 156 + 374 530	9. 184 + 275 459	10. 514 - 173 341

B. Solve the following problems.

	Show your work.
1. Sandy had 456 stamps. She gave 176 stamps to her little sister. How many stamps did she have left? There were __280__ stamps left.	
2. There were 127 girls and 98 boys playing in the school yard. How many children were playing altogether? __225__ children were playing.	

C. Circle the missing sign.

235 ◯ 169 = 404	(+) — =

Math Test: Addition and Subtraction

1. 67 + 6 73	2. 49 + 18 67	3. 31 - 18 13	4. 63 - 28 35
5. 134 + 537 671	6. 566 + 174 740	7. 975 - 347 628	8. 409 - 158 251
9. 7986 + 1647 9633	10. 45 967 + 12 563 58 530	11. 7542 - 2578 4964	12. 7472 - 5175 2297
13. 564 275 + 411 1250	14. 3756 1036 + 5014 9806	15. 89 407 - 69 758 19 649	16. 49 850 - 25 271 24 579
17. 65 331 33 986 + 13 745 113 062	18. 7356 1757 3611 + 7564 20 288	19. 66 108 - 11 977 54 131	20. 97 755 - 31 109 66 646

Number Correct
___
20

Name ____________________ **Addition**

4986 + 1780 6766	1785 + 4645 6430	7087 + 1788 8875	3082 + 5198 8280	7463 + 6117 13580	1337 + 7234 8571	1037 + 1235 2272
8523 + 1370 9893	2859 + 4752 7611	1763 + 2097 3860	3238 + 1634 4872	6379 + 2717 9096	8523 + 1470 9993	1523 + 2692 4215
4992 + 3512 8504	2611 + 1643 4254	1532 + 8623 10155	1754 + 2031 3785	8975 + 1478 10453	2034 + 3936 5970	Number Correct: ___ 20

Name ____________________ **Addition**

5673 + 5190 10863	2073 + 2750 4823	1734 + 3749 5483	7081 + 1747 8828	1267 + 7234 8501	7745 + 6512 14257	2043 + 1780 3823
3092 + 1906 4998	8893 + 1310 10203	2875 + 4832 7707	1163 + 2097 3260	3892 + 2450 6342	6124 + 2613 8737	8869 + 1312 10181
1651 + 5071 6722	4992 + 34601 39593	4611 + 1203 5814	1402 + 8784 10186	1035 + 3126 4161	8705 + 1368 10073	Number Correct: ___ 20

Name ____________________ **Subtraction**

7852 - 1378 6474	7785 - 4049 3736	7089 - 1788 5301	8082 - 4136 3946	7261 - 6017 1244	6370 - 2234 4136	5037 - 1563 3474
8523 - 5630 2893	2859 - 1787 1072	5061 - 2413 2648	8953 - 8624 329	3757 - 1718 2039	8523 - 4988 3535	2341 - 1558 783
4731 - 1312 3419	4521 - 1784 2737	9532 - 7431 2101	2044 - 1754 290	8789 - 1404 7385	5034 - 1936 3098	Number Correct: ___ 20

Name ____________________ **Subtraction**

5980 - 2397 3583	4723 - 2140 2583	3734 - 1749 1985	7634 - 1727 5907	7288 - 1256 6032	8740 - 5641 3099	2093 - 1675 418
3092 - 1452 1640	8893 - 1768 7125	6785 - 4832 1953	2007 - 1226 781	3892 - 1450 2442	6865 - 2618 4247	8162 - 1329 6833
5651 - 1081 4570	4902 - 1460 3442	5871 - 1203 4668	9578 - 1784 7794	9035 - 3156 5879	8705 - 3785 4920	Number Correct: ___ 20

th Test: Addition and Subtraction of Money

134.50	$173.50	$45.80	$56.20	$94.87	$97.71	$56.75
$ 70.45	- $ 90.72	- $21.06	- $10.09	- $32.38	- $53.33	- $13.37
$ 64.05	$ 82.78	$24.74	$46.11	$62.49	$44.38	$43.38

$8.01	$29.08	$86.50	$84.53	$51.74	$4.70	$5.70
- $3.30	+ $65.75	- $43.03	- $33.06	+ $32.38	+ $3.37	+ $9.38
$4.71	$94.83	$43.47	$51.47	$84.12	$8.07	$15.08

$51.70	$29.08	$24.83	$64.83	$54.72	$68.73	Number Correct:
$15.73	+ $13.56	- $11.09	- $11.37	- $27.03	- $22.04	___
$67.43	$42.64	$13.74	$53.46	$27.69	$46.69	20

th Test: Addition and Subtraction of Money

193.70	$50.82	$98.45	$154.76	$23.70	$91.75	$77.80
$ 96.79	- $16.09	- $13.38	- $ 13.40	- $11.66	- $34.38	- $63.90
$96.91	$34.73	$85.07	$141.36	$12.04	$57.37	$13.90

$89.00	$75.83	$5.70	$8.35	$81.23	$9.10	$7.12
$05.75	- $23.06	- $3.35	- $3.30	- $73.04	+ $9.38	+ $3.99
$94.75	$52.77	$2.35	$5.05	$8.19	$18.48	$11.11

$69.08	$23.83	$38.73	$32.99	$44.22	$38.70	Number Correct:
$10.53	- $11.37	- $27.03	+ $17.80	-$11.69	- $21.03	___
$79.61	$12.46	$11.70	$50.79	$32.53	$17.67	20

Name ____________________ **Multiplication Facts 1-10 #1**

2	5	4	8	6	10	7
X 5	X 8	X 2	X 1	X 3	x 4	X 6
10	40	8	8	18	40	42

6	9	6	3	7	9	9
X 4	X 5	X10	X 2	X 4	X 1	X 6
24	45	60	6	28	9	54

1	10	8	7	10	10	Number Correct:
X 6	X 7	X 8	X 5	X 2	X 9	___
6	70	64	35	20	90	20

Name ____________________ **Multiplication Facts 1-10 #2**

3	6	5	4	7	8	4
X 5	X 1	X 10	X 3	X 7	X 2	X 4
15	6	50	12	49	16	16

8	9	7	1	10	2	3
X 4	X 7	X 3	X 4	X 5	X 8	X 9
32	63	21	4	50	16	27

4	6	5	8	3	3	Number Correct:
X 7	X 6	X 5	X 3	X 3	X 6	___
28	36	25	24	9	18	20

Name ____________________ **Multiplication**

69	73	92	89	84	13	41
x 83	x 25	x 86	x 25	x 63	x 45	x 25
5727	1825	7912	2225	5292	585	1025

39	35	78	17	22	40	97
x 56	x 64	x 22	x 64	x 27	x 48	x 64
2184	2240	1716	1088	594	1920	6208

33	54	42	88	18	60	Number Correct:
x 14	x 17	x 56	x 17	x 35	x 32	___
462	918	2352	1496	630	1920	20

Name ____________________ **Multiplication**

74	49	37	60	28	49	19
x 93	x 35	x 56	x 73	x 15	x 25	x 35
6882	1715	2072	4380	420	1225	665

75	45	68	96	23	80	45
x 36	x 74	x 42	x 61	x 48	x 84	x 94
2700	3330	2856	5856	1104	6720	4230

63	82	38	55	80	82	Number Correct:
x 17	x 42	x 28	x 39	x 45	x 76	___
1071	3444	1064	2145	3600	6232	20

Math Test: Multiplication

Name ____________________

34	85	72	63	89	45	77
x 13	x 35	x 56	x 45	x 66	x 75	x 25
442	2975	4032	2835	5874	3375	1925

31	38	76	15	21	42	93
x 46	x 54	X32	x 84	x 17	x 48	x 64
1426	2052	2432	1260	357	2016	5952

23	64	41	98	15	50	Number Correct:
x 15	x 27	x 46	x 16	x 25	x 22	___
345	1728	1886	1568	375	1100	20

Math Test: Multiplication

Name ____________________

94	48	47	61	38	48	29
x 83	x 34	x 66	x 74	x 75	x 35	x 38
7802	1632	3102	4515	2850	1680	1102

85	35	64	86	13	81	49
x 37	x 77	x 62	x 62	x 98	x 44	x 54
3145	2695	3968	5332	1274	3564	2646

63	72	39	51	10	85	Number Correct:
x 10	x 46	x 78	x 59	x 46	x 26	___
630	3312	3042	3009	460	2210	20

Multiplication

1. 20 X 50 = 1000
2. 20 X 30 = 600
3. 50 X 50 = 2500
4. 50 X 40 = 2000
5. 40 X 10 = 400
6. 80 X 100 = 8000
7. 70 X 20 = 1400
8. 50 X 60 = 3000
9. 40 X 70 = 2800
10. 90 X 80 = 7200
11. 70 X 50 = 3500
12. 50 X 20 = 1000
13. 60 X 60 = 3600
14. 80 X 20 = 1600
15. 20 X 60 = 1200

Number Correct: /15

Multiplication

Name ______________

1. 10 X 10 = 100
2. 34 X 10 = 340
3. 6 X 10 = 60
4. 3 X 1000 = 3000
5. 69 X 10 = 690
6. 70 X 100 = 7000
7. 40 X 10 = 400
8. 22 X 10 = 220
9. 90 X 100 = 9000
10. 34 X 10 = 340
11. 4 X 1000 = 4000
12. 25 X 100 = 2500
13. 22 X 100 = 2200
14. 95 X 100 = 9500
15. 100 X 10 = 1000

Number Correct: /15

Math Test: Multiplication With Decimals

1.	2.	3.	4.	5.
2.3 x 5 = 11.5	3.8 x 8 = 30.4	6.9 x 3 = 20.7	1.2 x 4 = 4.8	7.3 x 5 = 36.5

6.	7.	8.	9.	10.
8.6 x 7 = 60.2	6.8 x 8 = 54.4	4.9 x 9 = 44.1	2.4 x 3 = 7.2	3.7 x 2 = 7.4

Number Correct: /10

Math Test: Multiplication With Decimals

1.	2.	3.	4.	5.
5.3 x 4.5 = 23.85	7.2 x 6.3 = 45.36	6.1 x 3.2 = 19.52	3.7 x 2.2 = 8.14	8.4 x 8.5 = 71.4

6.	7.	8.	9.	10.
9.3 x 2.1 = 19.53	1.8 x 4.9 = 8.82	2.5 x 8.7 = 21.75	7.4 x 1.6 = 11.84	8.2 x 3.3 = 27.06

Number Correct: /10

Name ____________________ **Division Facts 6-10 #1**

36 ÷ 4 =	21 ÷ 3 =	36 ÷ 9 =	27 ÷ 3 =	10 ÷ 2 =	30 ÷ 5 =	24 ÷ 4 =
9	7	4	9	5	6	6
24 ÷ 6 =	18 ÷ 9 =	32 ÷ 8 =	35 ÷ 7 =	32 ÷ 4 =	48 ÷ 6 =	49 ÷ 7 =
4	2	4	5	8	8	7
64 ÷ 8 =	63 ÷ 9 =	30 ÷ 6 =	40 ÷ 10 =	14 ÷ 7 =	27 ÷ 9 =	Number Correct:
8	7	5	4	2	3	___ / 20

Name ____________________ **Division Facts 6-10 #2**

27 ÷ 9 =	6 ÷ 6 =	20 ÷ 10 =	90 ÷ 10 =	42 ÷ 7 =	40 ÷ 5 =	30 ÷ 10 =
3	1	2	9	6	8	3
45 ÷ 5 =	16 ÷ 2 =	5 ÷ 5 =	8 ÷ 2 =	12 ÷ 4 =	1 ÷ 1 =	30 ÷ 5 =
9	8	1	4	3	1	6
80 ÷ 8 =	54 ÷ 6 =	36 ÷ 9 =	72 ÷ 9 =	28 ÷ 7 =	40 ÷ 10 =	Number Correct:
10	7	4	8	4	4	___ / 20

Math Test: Dividing Decimals

1) 2.6 ÷ 10 = .26	2) 5.4÷ 100 = .054
3) 716.9 ÷ 100 = 7.169	4) 62.9 ÷ 10 = 6.29
5) 426.9 ÷ 10 = 42.69	6) 985.2 ÷ 100 = 9.852
7) 0.3 ÷ 10 = .03	8) 62.4 ÷ 10 = 6.24
9) 593.1 ÷ 100 = 5.931	10) 204.6 ÷ 100 = 2.046

Number Correct: /10

Math Test: Dividing Decimals

1) 15.7÷ 100 = .157	2) 7.4 ÷ 100 = .074
3) 32.5 ÷ 10 = 3.25	4) 67.1 ÷ 10 = 6.71
5) 641.2 ÷ 100 = 6.421	6) 891.9 ÷ 100 = 8.919
7) 34.4 ÷ 10 = 3.44	8) 0.8 ÷ 10 = .08
9) 786.2 ÷ 10 = 78.62	10) 412.5 ÷ 100 = 4.125

Number Correct: /10

Math Test: Division- No Remainders

	Quotient	Problem
1)	24	5) 120
2)	89	6) 534
3)	191	5) 955
4)	70	6) 420
5)	297	3) 891
6)	95	8) 760
7)	266	3) 798
8)	46	9) 414
9)	90	8) 720
10)	186	5) 930

Math Test: Division- No Remainders

	Quotient	Problem
1)	38	7) 266
2)	225	4) 900
3)	490	2) 980
4)	114	3) 342
5)	79	7) 553
6)	112	8) 896
7)	78	6) 468
8)	46	9) 414
9)	166	5) 830
10)	136	3) 408

Math Test: Division- With Remainders

	Quotient	Problem
1)	124 r1	5) 621
2)	86 r2	9) 778
3)	88	4) 352
4)	35 r1	4) 141
5)	101 r3	6)609
6)	47 r2	8) 378
7)	132 r1	7) 925
8)	165 r1	3) 496
9)	60 r3	8) 483
10)	86 r6	7) 608

Math Test: Division- With Remainders

	Quotient	Problem
1)	106 r1	5) 531
2)	139 r1	6) 835
3)	115 r2	5) 577
4)	107 r1	3) 322
5)	27r7	9) 250
6)	26 r3	5) 133
7)	30 r2	6) 182
8)	39 r2	9)353
9)	87 r1	9) 784
10)	102 r1	3)307

Math Test: Division- No Remainders

	Quotient	Problem
1)	83	90) 7470
2)	2	50) 100
3)	25	30) 750
4)	170	20) 3400
5)	5	80)400
6)	5	30) 150
7)	55	90) 4950
8)	208	20)4160
9)	262	30) 7860
10)	12	70)840

Math Test: Division- With Remainders

	Quotient	Problem
1)	597r1	15) 8956
2)	46r76	96) 4492
3)	381r7	18) 6865
4)	805r3	11)8858
5)	100r57	65) 6557
6)	31r13	46)1439
7)	114r7	51) 5821
8)	79r6	16)1270
9)	10r10	81)820
10)	285r7	13)3712

Division with Decimals

	Quotient	Problem
1)	0.8	5) 4.0
2)	0.35	6) 2.1
3)	0.73	6) 4.38
4)	0.7	9) 6.3
5)	0.675	8) 5.400
6)	0.393	6) 2.358
7)	4.12	2) 8.24
8)	2.29	2) 4.58
9)	3.3	1.6) 5.28
10)	0 .9	7.8) 7.02
11)	0 .3	4.9) 1.47
12)	3.2	2.6) 8.32
13)	4.5	3.4) 15.3
14)	6.3	1.4) 8.82
15)	2.1	7.4) 15.54

Number Correct

/15

Math Test: Prime and Composite Numbers

Identify each number as **prime (P)** or **composite (C)**.

1. 8 C
2. 22 C
3. 54 C
4. 78 C
5. 72 C
6. 11 P
7. 45 C
8. 34 C
9. 121 C
10. 90 C
11. 53 P
12. 66 C
13. 73 P
14. 117 C
15. 99 C

Number Correct: /15

Math Test: Prime and Composite Numbers

Identify each number as **prime (P)** or **composite (C)**.

1. 88 C
2. 83 P
3. 4 C
4. 71 P
5. 29 P
6. 100 C
7. 94 C
8. 351 C
9. 67 P
10. 36 C
11. 5 P
12. 41 P
13. 17 P
14. 80 C
15. 98 C

Number Correct: /15

Math Test: Factors

Write all the factors for each given number below.

1.	14	1 2 7 14
2.	45	1 3 5 9 15 45
3.	88	1 2 4 8 11 22 44 88
4.	36	1 2 3 4 6 9 12 18 36
5.	24	1 2 3 4 6 8 12 24
6.	76	1 2 4 19 38 76
7.	60	1 2 3 4 5 6 10 12 15 20 30 60
8.	96	1 2 3 4 6 8 12 16 24 32 48 96
9.	44	1 2 4 11 22 44
10.	11	1 11
11.	50	1 2 5 10 25 50
12.	33	1 3 11 33
13.	55	1 5 11 55
14.	10	1 2 5 10
15.	66	1 2 3 6 11 22 33 66

Number Correct: /15

Math Test: Greatest Common Factor

Find the greatest common factor of each set of numbers.

1.	8 and 48	8
2.	50 and 12	2
3.	21 and 42	21
4.	36 and 12	12
5.	4 and 36	4
6.	72 and 32	8
7.	14 and 63	7
8.	6 and 12	6
9.	5 and 25	5
10.	90 and 30	30
11.	35 and 70	35
12.	50 and 95	5
13.	20 and 40	20
14.	32 and 48	16
15.	7 and 70	7

Number Correct: /15

Math Test: Least Common Multiple

Write all the factors for each given number below.

1.	8 and 20	40
2.	5 and 29	145
3.	4 and 7	28
4.	2, 5 and 10	10
5.	4 and 12	12
6.	6 and 17	102
7.	14 and 30	420
8.	6 and 12	12
9.	5 and 25	25
10.	8 and 12	24
11.	5 and 7	35
12.	3 and 8	24
13.	2 and 4	4
14.	6 and 11	66
15.	10 and 12	120

Number Correct: /15

Answer Pages

Page 23 Math Test: Percent of a Number

1. 32	2. 21	3. 108	4. 45	5. 27
6. 120	7. 51	8. 12	9. 26	10. 21

Page 24 to 26 Math Test: Number Sense Grade 4

11. B	12. A	13. C	14. B	15. B
16. B	17. C	18. A	19. A	20. B
21. A	22. A	23. C	24. A	25. B
26. C	27. B	28. C	29. C	30. B
31. C	32. C	33. B	34. C	35. C
36. C	37. A	38. A	39. C	40. C

Page 27 to 28 Math Test: Number Sense Grade 5

1. C	2. C	3. B	4. B	5. C
6. A	7. B	8. A	9. B	10. C
11. A	12. B	13. A	14. B	15. B
16. A	17. B	18. C	19. C	20. C

Page 29 to 30 Math Test: Number Sense Grade 6

1. B	2. A	3. C	4. A	5. C
6. C	7. A	8. B	9. A	10. B
11. A	12. A	13. B	14. A	15. A
16. B	17. C	18. C	19. C	20. B

Page 31 Math Test: Fractions

1. $\frac{7}{28}$	2. $\frac{28}{48}$	3. $\frac{9}{24}$	4. $\frac{42}{60}$	5. $\frac{8}{16}$
6. $\frac{56}{72}$	7. $\frac{20}{24}$	8. $\frac{4}{5}$	9. $\frac{6}{20}$	10. $\frac{24}{32}$
11. $\frac{2}{4}$	12. $\frac{4}{6}$	13. $\frac{6}{8}$	14. $\frac{3}{4}$	15. $\frac{6}{7}$

Page 32 Math Test: Fractions

A	1. answer varies	2. answer varies	3. answer varies
B	1. $\frac{21}{5}$	2. $\frac{17}{7}$	3. $\frac{11}{3}$
C	1. $\frac{2}{3}$	2. $\frac{1}{3}$	3. $\frac{1}{3}$
	4. $\frac{5}{7}$	5. $\frac{1}{2}$	6. $\frac{1}{8}$

D $\frac{1}{8}$, $\frac{2}{8}$, $\frac{5}{8}$ **E** $\frac{2}{5}$

Answer Pages

Page 33 Math Test: Decimals

1. 0.4 2. 0.35 3. 0.2 4. 0.67 5. 0.1
6. 0.12 7. 0.89 8. 0.3 9. < 10. <
11. < 12. = 13. > 14. > 15. >
16. > 17. 8.77, 8.12, 7.1, 3.1, 0.99 18. 3.9, 2.5, 1.80, 0.78,0 .2
19. 8.01, 9.302, 9.51, 9.6, 20. 2.7, 8.8, 9.71

Page 34 Math Test: Fractions

A 1. $\frac{3}{4}$ 2. $\frac{7}{5}$ 3. $\frac{5}{6}$ 4. $\frac{17}{30}$

B 1. $\frac{1}{2}$ 2. $\frac{1}{2}$ 3. $\frac{9}{14}$ 4. $\frac{1}{14}$

C 1. $\frac{8}{3}$ 2. $\frac{6}{4}$ 3. $\frac{2}{64}$ or $\frac{1}{32}$ 4. $\frac{4}{27}$

D 1. 12 2. $2\frac{4}{5}$ 3. $\frac{2}{3}$ 4. 3

Page 35 Math Test: Fractions, Percents, Decimals and Ratios

A.

Fraction	Denominator of 100	Percentage	Decimal	Ratio
1. $\frac{8}{20}$	$\frac{40}{100}$	40%	0.40	8:20 or 2:5
2. $\frac{7}{25}$	$\frac{28}{100}$	28%	0.28	7:25
3. $\frac{16}{10}$	$\frac{160}{100}$	160%	1.60	16:10 or 8:5
4. $\frac{3}{5}$	$\frac{60}{100}$	60%	0.60	3:5

B.
1. E 2. D 3. 50% 4. E 5. A 6. 8:30 or 4:15

Page 36 to 37 Math Test: How Much Money?

1. \$28.66 2. \$22.71 3. \$19.61 4. \$16.17 5. \$46.61
6. \$28.66 7. \$63.76 8. \$101.68 9. \$38.22 10. \$36.61

Page 38 Math Test: Money
1. 200 dimes 2. \$20, \$20, \$10, \$10, \$5 ,\$5, 10¢, 10¢,1¢,1¢ 3. \$148.90 4. \$7.20 5. \$58.80

Answer Pages

Page 40 to 41 Math Test: Order of Operations

1. 337	2. 24	3. 66	4. 45	5. 1100
6. 46	7. 82	8. 38	9. 4	10. 331
11. 324	12. 60	13. 18	14. 383	15. 186
16. 65	17. 17	18. 50	19. 20	20. 1871

Page 42 to 43 Math Test: Patterning and Algebra Grade 4

1. C	2. C	3. C	4. C	5. A
6. A	7. A	8. B	9. B	10. C
11. B	12. A	13. A	14. B	15. C
16. C	17. B	18. A	19. A	20. C

Page 44 to 45 Math Test: Patterning and Algebra Grade 5

1. B	2. A	3. C	4. B	5. B
6. B	7. C	8. B	9. B	10. B
11. A	12. C	13. A	14. A	15. B
16. A	17. C	18. A	19. A	20. A

Page 46 to 47 Math Test: Patterning and Algebra Grade 6

1. A	2. C	3. B	4. A	5. C
6. C	7. B	8. C	9. A	10. A
11. C	12. A	13. B	14. A	15. A
16. B	17. A	18. C	19. C	20. C

Page 49 Math Test: Naming Shapes

1. hexagon	2. pentagon	3. circle	4. square	5. parallelogram
6. rectangle	7. triangle	8. octagon	9. trapezoid	10. rhombus

Page 50 to 51 Math Test: Classifying and Sorting Shapes

1. pentagon	2. triangle	3. hexagon and pentagon	4. hexagon trapezoid and square	5. all but the circle
6. rectangle and square	7. circle	8. octagon and hexagon	9. triangle, hexagon and square	10. square, rhombus and parallelogram

Page 52 Math Test: 3D Figures

1. F	2. E	3. B	4. A	5. D	6. C

Answer Pages

Page 53 Math Test: 3D Figures

1. Cylinder : Faces 2 edges 2 vertices 0
2. Sphere : Faces 0 edges 0 vertices 0
3. Rectangular Prism : Faces 6 edges 12 vertices 8
4. Cone : Faces 1 edges 1 vertices 0
5. Pyramid : Faces 5 edges 8 vertices 5
6. Cube : Faces 6 edges 12 vertices 8

Page 54 Math Test: Symmetry

1. 2	2. 1	3. 0	4. 0	5. 1
6. 1	7. 1	8. 0	9. 1	10. 0
11. 1	12. 1	13. 0	14. 1	15. 0

Page 55 Math Test: Angles

A 1. obtuse 2. right angle 3. acute

B 1. acute 2. right 3. obtuse 4. obtuse 5. right 6. acute

C 30° **D** 120° **E** B

Page 56 Math Test: Classifying Angles

1. acute	2. acute	3. straight	4. acute	5. right
6. acute	7. right	8. obtuse	9. obtuse	10. straight

Page 57 Math Test: Angles

1. 59 °, isosceles	2. 55 °, scalene	3. 60 °, scalene	4. 40 °, scalene	5. 19 °, scalene
6. 60 °, equilateral	7. 70 °, scalene	8. 25 °, scalene	9. 20 °, isosceles	10. 15 °, scalene

Page 58 to 59 Math Test: Geometry Grade 4

1. A	2. C	3. A	4. C	5. B
6. C	7. A	8. C	9. A	10. A
11. A	12. C	13. B	14. C	15. B
16. B	17. C	18. A	19. A	20. B

Page 60 to 61 Math Test: Geometry Grade 5

1. A	2. B	3. A	4. B	5. A
6. B	7. B	8. B	9. A	10. B
11. A	12. C	13. B	14. B	15. A
16. B	17. A	18. B	19. C	20. B

Page 62 to 63 Math Test: Geometry Grade 6

1. C	2. A	3. B	4. A	5. B
6. C	7. C	8. B	9. A	10. C
11. C	12. B	13. A	14. B	15. B
16. A	17. A	18. A	19. B	20. B

Answer Pages

Page _64_ Math Test: Ordered Pairs

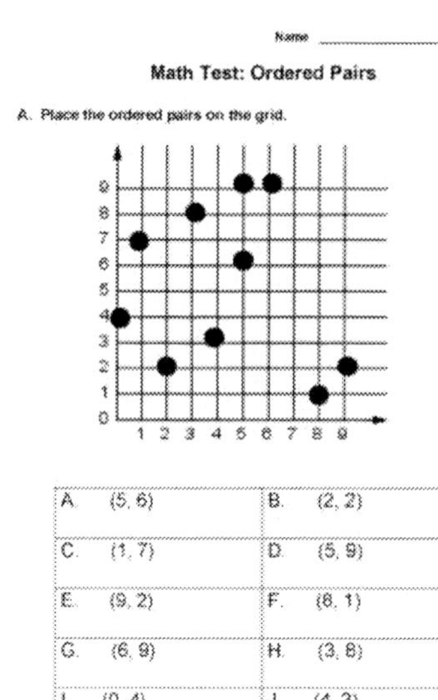

Name ________________

Math Test: Ordered Pairs

A. Place the ordered pairs on the grid.

A. (5, 6)	B. (2, 2)
C. (1, 7)	D. (5, 9)
E. (9, 2)	F. (8, 1)
G. (6, 9)	H. (3, 8)
I. (0, 4)	J. (4, 3)

Number Correct /10

Chalkboard Publishing © 2006 64

Page _65_ Math Test: Ordered Pairs

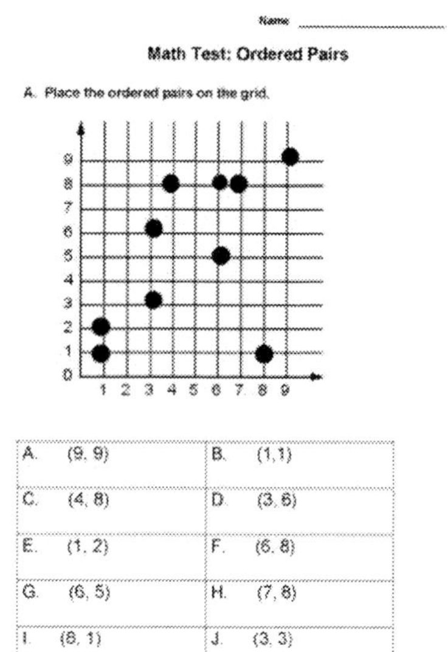

Name ________________

Math Test: Ordered Pairs

A. Place the ordered pairs on the grid.

A. (9, 9)	B. (1,1)
C. (4, 8)	D. (3, 6)
E. (1, 2)	F. (6, 8)
G. (6, 5)	H. (7, 8)
I. (8, 1)	J. (3, 3)

Chalkboard Publishing © 2006 65

Page _67_ Math Test: Telling Time

1. 8:05
2. 7:10
3. 5:25
4. 10:15
5. 8:35
6. 8:55
7. 12:30
8. 6:45
9. 11:05
10. 2:15
11. 3:15
12. 5:35

Page _68_ Math Test: Telling Time

1. 3:07
2. 8:30
3. 8:56
4. 8:14
5. 5:25
6. 2:15
7. 4:32
8. 8:00
9. 5:45
10. 3:50
11. 6:05
12. 11:41

Page _69_ Math Test: Calculate the Elapsed Time

1. 4:54 a.m.
2. 3:25 p.m.
3. 11:34 p.m.
4. 12:05 a.m.
5. 8:35 p.m.
6. 11:44 p.m.
7. 10:49 a.m.
8. 12:07 a.m.
9. 1:59 a.m.
10. 9:30 p.m.

Page _70_ Math Test: Time

1. B
2. C
3. C
4. B
5. A
6. A
7. B
8. A
9. B
10. B

Answer Pages

Page 71 Relating Units of Measurement

A.

1. 300	2. 300	3. 4000	4. 500	5. 6	6. 220
7. 7	8. 400	9. 1400	10. 8	11. 5	12. 800
13. 2	14. 6000	15. 3000	16. 100	17. 900	18. 1

B.

1. 700	2. 100	3. 5000	4. 9	5. 2000	6. 90
7. 200	8. 3	9. 7600	10. 7	11. 450	12. 90 000
13. 500	14. 8	15. 200	16. 4	17. 100	18. 4

Page 72 Math Test: Units of Measure

1. metres	2. centimetres	3. kilometres	4. metres	5. millimetres
6. metres or decametres	7. centimetres	8. grams	9. seconds	10. millimetres
11. milliliters	12. tonnes	13. litres	14. hours	15. degrees celcius

Page 73 Math Test: Perimeter

A 1. 16 2. 14 3. 14 4. 18

B Figure 2 has the longest perimeter

Page 74 Math Test: Area

A 1. 9 2. 10 3. 10 4. 16

B Both figures have the same area.

Page 75 Math Test: Perimeter and Area

1. 18 units, 16 square units	2. 18 units, 14 square units	3. 24 units, 13 square units
4. 24 units, 15 square units	5. 18 units, 15 square units	6. 22 units, 14 square units
7. 22 units, 15 square units	8. 22 units, 15 square units	9. 18 units, 20 square units

Page 76 Math Test: Perimeter

1. 38	2. 36	3. 20	4. 26	5. 24
6. 15	7. 24	8. 30	9. 24	10. 24

Page 77 Math Test: Perimeter and Area

1. p= 26 m 2. p= 21 m 3. p= 14 m 4. p= 18 m 5. p= 20 m

Page 78 Math Test: Perimeter and Area

1. 42 m 104 m^2	2. 52 m 153 m^2	3. 50 m 126 m^2	4. 44 m 96 m^2	5. 28 m 28 m^2	6. 16 cm 14 cm^2

Answer Pages

Page 79 Math Test: Area

1. 40.5 cm^2	2. 50 cm^2	3. 30.4 cm^2	4. 21 cm^2
5. 36 cm^2	6. 7.5 cm^2	7. 17.5 cm^2	8. 168 cm^2

Page 81 Math Test: Volume

1. 216 cm^3 2. 324 cm^3 3. 72 cm^3 4. 54 mm^3 5. 72 cm^3 6. 343 cm^3

Page 82 Math Test: Measurement Grade 4

1. B	2. A	3. A	4. C	5. C
6. B	7. A	8. A	9. A	10. B

Page 83 Math Test: Measurement Grade 4

11. A	12. A	13. B	14. C	15. C
16. B	17. C	18. B	19. B	20. B

Page 84 Math Test: Measurement Grade 5

1. C	2. A	3. A	4. A	5. C
6. A	7. C	8. A	9.A	10. B

Page 85 Math Test: Measurement Grade 5

11. A	12. C	13. A	14. B	15. B
16. A	17. B	18. C	19. C	20.B

Page 86 Math Test: Measurement Grade 6

1. C	2. B	3. A	4. A	5. C
6. A	7. B	8. B	9. A	10. B

Page 87 Math Test: Measurement Grade 6

11. C	12. A	13. A	14. A	15. C
16. B	17. B	18. B	19. A	20. B

Answer Pages

Page 89 Data Management #1 and #2

Set of Data	Mean	Range	Median	Mode
1. 28, 21, 3, 10, 6, 9, and 21	14	25	10	21
2. 24, 3, 17, 23, 19, 19, and 7	16	21	19	19
3. 13, 4,9 23, 24, and 23	16	20	18	23
4. 12, 21, 12, 7,and 8	12	14	12	12
5. 7, 19, 28, 8, and 28	18	21	19	28

Set of Data	Mean	Range	Median	Mode
1. 5, 25, 14, 25, 25, 2, and 2	14	23	14	25
2. 7, 6, 2, 6, 28, 4, and 10	9	24	6	6
3. 15, 27, 28, 5, and 5	16	23	15	5
4. 22, 25, 17, 22, and 4	18	21	22	22
5. 7, 28, 7, 6, and 7	11	21	7	7

Page 90 Math Test: Data Management #3 and #4

Set of Data	Mean	Range	Median	Mode
1. 16, 17, 15, 16, and 11	15	6	16	16
2. 27, 15, 15, 2, and 26	17	25	15	15
3. 3, 3, 6, 29, 21, 20, and 2	12	27	6	3
4. 14, 11, 14, 19, and 22	16	11	14	14
5. 5, 8, 28, 5, and 19	13	23	8	5

Data Management #4

Set of Data	Mean	Range	Median	Mode
1. 18, 12, 11, 2, and 17	12	16	12	no mode
2. 15, 25, 25, 10, and 25	20	15	25	25
3. 6, 28, 10, 10, and 11	13	22	10	10
4. 24, 24, 2, 22, and 23	19	22	23	24
5. 7, 17, 26, 17, 3,	14	23	17	17

Page 91 Math Test: Data Management

1. 41
2. 5
3. 24-8 or 16
4. 8
5. 63
6. 17
7. 68
8. 22
9. 11
10. 18.6

Answer Pages

Page 92 Math Test: Data Management
Answers will vary. Some examples are: Class A collected the most. Class A had a decreasing trend. Class B had an increasing trend. Class A and Class B collected the same amount in week 3.

Page 93 Math Test: Constructing Bar Graphs

A. Reading 9 Math 7 Art 7 Science 4

B. Answers will vary: Some examples are: 27 people voted altogether. Reading was the most popular. Science was the least popular. Math and Art were the same.

Page 94 Math Test: Data Management
1.

Football	Baseball	Basketball	Hockey	Tennis
𝍸 𝍸 𝍸 𝍸 𝍸	𝍸 𝍸 𝍸 𝍸 𝍸 𝍸 𝍸 𝍸 𝍸 𝍸 𝍸 𝍸 𝍸 𝍸 𝍸	𝍸 𝍸 𝍸 𝍸	𝍸 𝍸 𝍸 𝍸 𝍸 𝍸 𝍸 𝍸 𝍸 𝍸 𝍸 𝍸 𝍸 𝍸	𝍸 𝍸

3. baseball
4. tennis
5. 200

Page 95 Math Test: Data Management

1. vanilla
2. chocolate chip
3. 72 students
4. 42 students
5. 138 students
6. 30 students
7. 24 students
8. 11:4
9. 12:5
10. 192 people

Page 96 Math Test: Data Management

B. Answers will vary. Some examples are: Game one had the fewest people, Game four had the most people, 150 more people attended game two than game one, 1100 people attended the four hockey games

C. Line graphs show a change over time. By using a line graph, we can see how the hockey game attendance changed from game to game.

Page 97 Math Test: Data Management

1. Game #1
2. Game #4
3. Game #4
4. Games #2 and #4
5. 1400 people
6. Game #2
7. Game #4

Math Test Certificates

Great Job!

____

Math Expert!

You did it!

____

Great Work!